T0192200

Communications
in Computer and Information Science   2008

Rationale

The CCIS series is devoted to the publication of proceedings of computer science conferences. Its aim is to efficiently disseminate original research results in informatics in printed and electronic form. While the focus is on publication of peer-reviewed full papers presenting mature work, inclusion of reviewed short papers reporting on work in progress is welcome, too. Besides globally relevant meetings with internationally representative program committees guaranteeing a strict peer-reviewing and paper selection process, conferences run by societies or of high regional or national relevance are also considered for publication.

Topics

The topical scope of CCIS spans the entire spectrum of informatics ranging from foundational topics in the theory of computing to information and communications science and technology and a broad variety of interdisciplinary application fields.

Information for Volume Editors and Authors

Publication in CCIS is free of charge. No royalties are paid, however, we offer registered conference participants temporary free access to the online version of the conference proceedings on SpringerLink (http://link.springer.com) by means of an http referrer from the conference website and/or a number of complimentary printed copies, as specified in the official acceptance email of the event.

CCIS proceedings can be published in time for distribution at conferences or as post-proceedings, and delivered in the form of printed books and/or electronically as USBs and/or e-content licenses for accessing proceedings at SpringerLink. Furthermore, CCIS proceedings are included in the CCIS electronic book series hosted in the SpringerLink digital library at http://link.springer.com/bookseries/7899. Conferences publishing in CCIS are allowed to use Online Conference Service (OCS) for managing the whole proceedings lifecycle (from submission and reviewing to preparing for publication) free of charge.

Publication process

The language of publication is exclusively English. Authors publishing in CCIS have to sign the Springer CCIS copyright transfer form, however, they are free to use their material published in CCIS for substantially changed, more elaborate subsequent publications elsewhere. For the preparation of the camera-ready papers/files, authors have to strictly adhere to the Springer CCIS Authors' Instructions and are strongly encouraged to use the CCIS LaTeX style files or templates.

Abstracting/Indexing

CCIS is abstracted/indexed in DBLP, Google Scholar, EI-Compendex, Mathematical Reviews, SCImago, Scopus. CCIS volumes are also submitted for the inclusion in ISI Proceedings.

How to start

To start the evaluation of your proposal for inclusion in the CCIS series, please send an e-mail to ccis@springer.com.

Suparna Dhar · Sanjay Goswami ·
U. Dinesh Kumar · Indranil Bose ·
Rameshwar Dubey · Chandan Mazumdar
Editors

AGC 2023

First Analytics Global Conference, AGC 2023
Kolkata, India, April 28–29, 2023
Proceedings

Editors
Suparna Dhar 🆔
NSHM Institute of Computing & Analytics,
NSHM Knowledge Campus Kolkata
Kolkata, India

U. Dinesh Kumar
Indian Institute of Management Bangalore
Bengaluru, India

Rameshwar Dubey 🆔
Liverpool Business School
Liverpool John Moores University
Liverpool, UK

Sanjay Goswami 🆔
NSHM Knowledge Campus
Kolkata, India

Indranil Bose 🆔
Indian Institute of Management Ahmedabad
Ahmedabad, Gujarat, India

Chandan Mazumdar
Department of Computer Science
and Engineering
Jadavpur University
Kolkata, West Bengal, India

ISSN 1865-0929   ISSN 1865-0937 (electronic)
Communications in Computer and Information Science
ISBN 978-3-031-50814-1   ISBN 978-3-031-50815-8 (eBook)
https://doi.org/10.1007/978-3-031-50815-8

This Springer imprint is published by the registered company Springer Nature Switzerland AG
The registered company address is: Gewerbestrasse 11, 6330 Cham, Switzerland

Paper in this product is recyclable.

Preface

Analytics Society of India Kolkata Chapter created by NSHM Institute of Computing & Analytics presented a two-day conference on the 28th and 29th of April, 2023 at CII, Salt Lake, Kolkata. The conference was inaugurated by Rajeev Kumar, Principal Secretary of Information Technology & Electronics, West Bengal, India, Dinesh Kumar from IIM Bangalore, Indranil Bose from IIM Ahmedabad, and Ambarish Dasgupta, Senior Partner, Intueri Consulting LLP along with Cecil Antony, Chief Mentor of NSHM Knowledge Campus, and the Chief Convener of the conference Suparna Dhar. The objective of the conference was industry-academia interfacing in the domain of machine learning and artificial intelligence. More than 40 eminent speakers from industry and academia enlightened the audience over the two days of the conference. The speakers represented IIM Ahmedabad, IIM Bangalore, IIM Calcutta, Indian Statistical Institute (ISI), IIT Delhi, Liverpool Business School, Indian Institute of Engineering Science and Technology (IIEST) Shibpur, Jadavpur University, Calcutta University, Indian Institute of Science Education and Research (IISER) Kolkata, Indian Institute of Foreign Trade (IIFT), Amazon, Google, IBM, Wipro, TCS, RS Software, Business Brio, PwC, L&T, Adani Group, Intueri Consulting, Achievex Solutions, Ingram Micro, Ananda Bazar Patrika, NASSCOM, Neuropixel.ai, Kolkata Police, and the Information Technology Department of West Bengal. The discussions spanned different aspects of analytics, right from the fundamental statistical theories that forms the base upon, to the technological aspects of data management, and analysis, and applications of analytics across industry sectors. This conference marked the beginning of an analytics culture in the region. 300+ participants including students, scholars, faculty members, and industry professionals attended the conference.

The conference received a good response from the research community. The topics were organized into two broad classes, namely, (1) applications of analytics in business, and (2) machine learning, deep learning and text analytics. Applications of analytics in business covered business process development with analytics; applications of analytics in cybersecurity, law and order, and governance; application of analytics in disaster management; application of analytics for societal development, education, health, etc.; and case studies on the application of analytics. Machine learning, deep learning and text analytics included data collection and preprocessing for analytics and AI/ML; data visualization for analytics; machine learning; deep learning; social media analytics; and text, image, and video analytics.

A total of 36 papers were submitted through the Springer EquinOCS system. The papers were subjected to a stringent peer review process, where each paper was subjected to three peer reviews. The revised papers were reviewed to check compliance. Scholars and practitioners presented research papers accepted for the conference. The paper presentation was judged by a highly accomplished jury. Eleven papers selected after the paper presentation were accepted for publication in the proceedings. The eleven papers include four papers on the application of analytics in business and seven papers on machine learning, deep learning and text analytics.

We express our gratitude to Springer for granting us the privilege to publish the conference proceedings within the CCIS series, as well as for their invaluable guidance throughout the publication process. Our appreciation extends to the authors for placing their trust in us, the reviewers for their punctual and thorough evaluations, and the Organizing Committee members, volunteers, and participants whose collective efforts ensured the conference's success. Lastly, we extend our thanks to our sponsors and mentors for their trust and unwavering support in our conference organization endeavors. The conference series aims to motivate students and young scholars to engage in quality research and impactful projects in analytics that will help build a strong analytics ecosystem in the region. We hope that the rich experience gained in making AGC 2023 a success will guide us in making future editions of the Analytics Global Conference more enduring and impactful.

November 2023

Suparna Dhar
Sanjay Goswami
U. Dinesh Kumar
Indranil Bose
Rameshwar Dubey
Chandan Mazumdar

Organization

Program Committee Chairs

Bose, Indranil	IIM Ahmedabad, India
Dhar, Suparna	NSHM Knowledge Campus Kolkata, India
Dubey, Rameshwar	Montpellier Business School, France; School of Management Liverpool Business School, UK
Goswami, Sanjay	NSHM Institute of Computing & Analytics, NSHM Knowledge Campus Kolkata, India
Mazumdar, Chandan	Jadavpur University, Kolkata, India
U. Dinesh Kumar	IIM Bangalore, India

Program Committee Members

Aslam, Humaira	NSHM Institute of Computing & Analytics, NSHM Knowledge Campus Kolkata, India
Banerjee, Soma	GBSM Consulting Private Limited (Business Brio), India
Bose, Indranil	IIM Ahmedabad, India
Chakraborty, Dipayan	Tata Consultancy Services, India
Chatterjee, Partha Sarathi	NSHM Knowledge Campus Kolkata, India
Das, Swapan	NSHM Knowledge Campus Kolkata, India
Datta, Pratyay Ranjan	NSHM Institute of Computing & Analytics, India
Debnath, Shankhya	NSHM Knowledge Campus Kolkata, India
Dhar, Suparna	NSHM Knowledge Campus Kolkata, India
Dubey, Rameshwar	Montpellier Business School, France; School of Management Liverpool Business School, UK
Gaur, Rajat	Ernst & Young LLP, India
Goswami, Sanjay	NSHM Knowledge Campus Kolkata, India
Kar, Ashutosh	NSHM Knowledge Campus Kolkata, India
Karn, Prashant	NSHM Knowledge Campus Kolkata, India
Mallik, Majid	NSHM Knowledge Campus Kolkata, India
Mandal, Sourav	XIM University, India
Mazumdar, Chandan	Jadavpur University, Kolkata, India
Mazunder, Sangita	NSHM Knowledge Campus Kolkata, India
Misra, Sumit	RS Software India Limited, India
Mukherjee, Amritendu	NeuroPixel.AI Labs, India

Mukherjee, Anik   IIM Calcutta, India
Pal, Sanjay Kumar   NAHM Knowledge Campus Kolkata, India
Paul, Madhurima   NAHM Knowledge Campus Kolkata, India
Paul, Poulomi   Tata Consultancy Services Ltd., UK
Roy, Moumita   NSHM Knowledge Campus Kolkata, India
Sarkar, Dhrubasish   Supreme Knowledge Foundation, Hoogly, India
Sarkar, Krishnendu   NSHM Knowledge Campus Kolkata, India
Sharib, Md   NSHM Knowledge Campus Kolkata, India
U. Dinesh Kumar   IIM Bangalore, India

Reviewers

Aslam, Humaira   NSHM Knowledge Campus Kolkata, India
Chatterjee, Partha Sarathi   NSHM Knowledge Campus Kolkata, India
Das, Swapan   NSHM Knowledge Campus Kolkata, India
Datta, Pratyay Ranjan   NSHM Institute of Computing & Analytics, India
Debnath, Shankhya   The Regional Institute of Printing Technology,
                                        Kolkata, India
Gaur, Rajat   Ernst & Young LLP, India
Goswami, Sanjay   NSHM Knowledge Campus Kolkata, India
Kar, Ashutosh   NSHM Knowledge Campus Kolkata, India
Karn, Prashant   NSHM Knowledge Campus Kolkata, India
Mallik, Majid   NSHM Knowledge Campus Kolkata, India
Mandal, Sourav   XIM University, India
Mazunder, Sangita   NSHM Knowledge Campus Kolkata, India
Misra, Sumit   RS Software India Limited, India
Mukherjee, Amritendu   NeuroPixel.AI Labs, India
Mukherjee, Anik   IIM Calcutta, India
Pal, Sanjay Kumar   NAHM Knowledge Campus Kolkata, India
Paul, Madhurima   NAHM Knowledge Campus Kolkata, India
Paul, Poulomi   Tata Consultancy Services Ltd., UK
Roy, Moumita   NSHM Knowledge Campus Kolkata, India
Sarkar, Dhrubasish   Supreme Knowledge Foundation, Hoogly, India
Sarkar, Krishnendu   NSHM Knowledge Campus Kolkata, India
Sharib, Md   NSHM Knowledge Campus Kolkata, India

Contents

Applications of Analytics in Business

Efficiency and Benchmarking Using DEA and Tessellation in Retail Stores 3
 Soma Banerjee, Riddhiman Syed, Ayan Chakraborty, Gautam Banerjee,
 and Abhishek Banerjee

A Machine Learning Framework Developed by Leveraging
the Topological Pattern Analysis of Bitcoin Addresses in Blockchain
Network for Ransomware Identification 23
 Dipayan Chakraborty, Sangita Mazumder, and Ashutosh Kar

A Taxonomy and Survey of Software Bill of Materials (SBOM) Generation
Approaches 40
 Vandana Verma Sehgal and P. S. Ambili

Find Your Donor (FYD): An Algorithmic Approach Towards Empowering
Lives and Innovating Healthcare 52
 Tamoleen Ray

Machine Learning, Deep Learning and Text Analytics

Deep Learning-Based Multiple Detection Techniques of Covid-19 Disease
From Chest X-Ray Images Using Advanced Image Processing Methods
and Transfer Learning 65
 Arif Hussain, Rohini Basak, and Sourav Mandal

Classification Model to Predict the Outcome of an IPL Match 83
 Poulomi Paul, Pratyay Ranjan Datta, and Ashutosh Kar

Knowledge Graph-Based Evaluation Metric for Conversational AI
Systems: A Step Towards Quantifying Semantic Textual Similarity 112
 Rajat Gaur and Ankit Dwivedi

Are We Nearing Singularity? A Study of Language Capabilities of ChatGPT ... 125
 Suparna Dhar and Indranil Bose

Thresholding Techniques Comparsion on Grayscale Image Segmentation 136
 Vinay Kumar Nassa and Ganesh S. Wedpathak

Automated Pneumonia Diagnosis from Chest X-rays Using Deep Learning
Techniques 147
  Pratyay Ranjan Datta and Moumita Roy

Detecting Emotional Impact on Young Minds Based on Web Page Text
Classification Using Data Analytics and Machine Learning 170
  Arjama Dutta, Tuhin Kumar Mondal, Shakshi Singh, and Saikat Dutta

Author Index 183

Applications of Analytics in Business

Efficiency and Benchmarking Using DEA and Tessellation in Retail Stores

Soma Banerjee[(⊠)], Riddhiman Syed, Ayan Chakraborty, Gautam Banerjee, and Abhishek Banerjee

GBSM Consulting Pvt Ltd. (Business Brio), Kolkata, India
{soma,gautam}@businessbrio.com, {riddhiman,ayan, abhishek}@businessbrio.in

Abstract. In today's highly competitive in-store retail industry, it is important to compare performance of stores. Techniques to evaluate a store's productivity and its relative efficiency involves evaluating many factors - such as sales turnover, sales per floor space, employee productivity, customer traffic, catchment area, gross margin, among other factors. Generally, store efficiency is evaluated by considering two metrics at a time, like sales per floor space or customers per employee.

This paper used Data Envelopment Analysis (DEA) - a non-parametric linear programming procedure which combined multiple inputs and outputs, to get the overall performance in terms of an efficiency score, which was then used to compare retail stores. For input consideration, the paper also used Voronoi tessellation method and mapped it to census data to get the target population i.e. serviceable addressable market (SAM) by identifying catchment area for each store. The specific case pertains to 30 stores in the state of Ohio in North America for a multi-brand retail chain for the purpose of the paper. This paper presents the application of these two methods seamlessly for the business benefit of retail industry for the first time in one single use case. The findings apply to companies that are trying to compare the efficiency of retail stores and can be further extended with competitor information on product and promotions from market surveys. This approach is also applicable to other verticals of healthcare and education to compare performance of healthcare units or clinics or institutes with relevant input and output parameters.

Keywords: Efficiency · Tessellation · Data Envelopment Analysis (DEA) · retail analytics

1 Introduction

Measuring operational efficiency is crucial for the success of a business. It helps decision makers to identify areas for improvement, increase productivity, improve customer experience and gain a competitive advantage. By streamlining processes, eliminating waste, and improving productivity, businesses can produce more output with the same resources or produce the same output with fewer resources.

Specifically in the retail sector, it is important for several reasons, including:

© The Author(s), under exclusive license to Springer Nature Switzerland AG 2024
S. Dhar et al. (Eds.): AGC 2023, CCIS 2008, pp. 3–22, 2024.
https://doi.org/10.1007/978-3-031-50815-8_1

- Improved Productivity: Measuring operational efficiency helps retailers identify areas where they can streamline processes, eliminate waste, and increase productivity. This can help to increase the output of goods and services, reduce costs and improve profitability.
- Better Customer Experience: By measuring operational efficiency, retailers can assess the speed, accuracy and quality of customer service, and identify areas for improvement. This can help to enhance the customer experience, increase customer satisfaction and build customer loyalty.
- Inventory Management: Measuring operational efficiency can help retailers to better manage their inventory levels and reduce waste. This can help to reduce the cost of carrying inventory, improve the accuracy of inventory counts and reduce the risk of stock shortages.

zip	lat	lng	city	state_id	population
43001	40.08794	-82.61289	Alexandria	OH	2507.0
43002	40.05982	-83.17305	Amlin	OH	3431.0
43003	40.41017	-82.96866	Ashley	OH	2936.0
43004	40.01664	-82.80025	Blacklick	OH	27972.0
43005	40.28241	-82.26728	Bladensburg	OH	392.0
...	...	...	...	...	...
45895	40.56991	-84.15197	Wapakoneta	OH	17394.0
45896	40.60667	-83.93775	Waynesfield	OH	1917.0
45897	40.83434	-83.65676	Williamstown	OH	91.0
45898	40.73617	-84.76429	Willshire	OH	914.0
45899	40.80088	-84.77491	Wren	OH	184.0

Fig. 1. US census data by zipcode

- Supply Chain Management: Retailers can use operational efficiency metrics to monitor the performance of their supply chain partners and identify areas for improvement. This can help to improve the reliability and speed of the supply chain, reduce costs and increase profitability.
- However, there are certain key challenges in measuring efficiency, including but not limited to:
- Data Availability: One of the biggest challenges in measuring operational efficiency is the availability and quality of data. Retailers may have difficulty collecting accurate and comprehensive data from different departments and systems, making it challenging to make informed decisions.
- Complex Processes: Retail operations can be complex, involving multiple departments, systems and processes. Measuring the efficiency of these processes can be

challenging, as it may require detailed analysis and a thorough understanding of the processes involved.

- Subjectivity: Measuring operational efficiency can be subjective, as different individuals may have different interpretations of what constitutes efficiency. This can make it difficult to develop meaningful metrics and compare performance across different retail operations.
- Resistance to Change: Change can be challenging, and many retailers may resist implementing changes that are necessary for improving operational efficiency. This can make it difficult to implement new processes, technologies or systems that are designed to increase efficiency.
- Lack of Standardization: There is a lack of standardization in the retail sector when it comes to measuring operational efficiency. This can make it challenging to compare performance across different retailers and to make informed decisions based on performance metrics.

This paper presents a case study of a combined application of computational geometry (Voronoi Tessellation) and economic modeling (Data Envelopment Analysis) techniques to determine efficiency of retail stores based on a combination of demographic and operation's data, keeping in mind the aforementioned challenges, and seeks to present decision makers with insights for determining efficient and inefficient units in their business, and therefore take remedial measures.

The case study does not present comparative analysis of different methods of Tessellation (while Voronoi variant is being used) nor different methods of DEA (CRS Output oriented method is used) as available in theories but presents the application of these two methods seamlessly for the business benefit of retail industry for the first time in one single use case.

The paper is organized into various sections, with Sect. 2 discussing the problem, Sect. 3 discussing the methods used. Implementation is discussed in Sect. 4 along with outcomes under Sect. 5, discussion in Sect. 6 followed by conclusion in Sect. 7.

2 Problem Statement

2.1 Background

As mentioned before, operational efficiency is a key metric in estimating performance of business units [5]. However, due to challenges such as subjectivity and lack of standardization, it becomes difficult to perform any benchmarking for business units and thereby figure out slacks, if any, in the performance of the units.

Ordinarily for a retail chain, efficiency may be understood as a ratio of the outputs of a business process to the inputs provided [8]. Inputs may be defined as resources such as capital, labor, infrastructure (floor space), target population with identified catchment area, number of product categories and SKUs, whereas outputs may be defined as revenue (sales value), acquired customers (unique), number of transactions, volume of items sold etc.

However, when multiple inputs and outputs are involved (as is the case in any practical business scenario), estimating efficiency becomes a complicated process. This in turn prevents management from identifying underperforming business units and take remedial measures or redistribute resources [9].

The specific case here refers to 30 stores in the state of Ohio in North America for a multi-brand retail chain for the purpose of the paper. The original project included other states and hence more stores. Regional census data from public data sources along with operations' metrics from the stores have been factored in as well. Certain inputs and output attributes have been prioritized with the help of correlation along with analytical hierarchical process (AHP) method. From this, efficiencies of stores have been identified and analyzed, keeping in mind current business constraints and regional target groups. Multiple algorithms have been deployed for the above analysis, including but not limited to Voronoi Tessellation, Data Envelopment Analysis among other methods.

2.2 Data Overview

2.2.1 Description

A sample of 30 stores spread across the state of Ohio was used in the study. The Population metric was taken from the regional census (public) data available which was mapped to the tiled regions obtained from the tessellation algorithm. This Population metric, along with the Floor_Space data, was prioritized using relevant methods as inputs and output variables, namely Sales (value) and Customers (unique) were used in the DEA model.

2.2.2 Attributes

- *zip*: United States zipcodes available in census data
- *lat*: Latitudes of zipcodes available in census data
- *lng*: Longitudes of zipcodes available in census data
- *city*: Zipcode city available in census data
- *state_id*: Zipcode state available in census data
- *population*: Population in zipcode location available in census data
- *Store_ID*: Unique identifier for each store
- *addr:city*: Store location city
- *addr:state*: Store location state
- *geometry*: Store location coordinates
- *latitude*: Latitude of store location
- *longitude*: Longitude of store location
- *Tile_ID*: Unique identifier for each tile generated via Voronoi Tessellation
- *DMU_Name*: Descriptive name of each decision-making unit i.e. store
- *SAM*: Serviceable addressable market enclosed within the catchment area for each DMU
- *Density*: The density of the SAM enclosed within the catchment area for each DMU
- *Tile_Region*: Boundary information for each tiled region
- *Floor_Space*: The available shopping area for a particular store
- *Sales*: Daily revenue for each store
- *Customers*: Daily customers per store

2.2.3 Units and Scales

- *Population* --> 1: 100000 people
- *Density* --> 1: 10000 people/sq. mile
- *Floor_Space* --> 1: 10000 sq. ft.
- *Sales* --> 1: 10000 dollars per day
- *Customers* --> 1: 100 per day

3 Methods and Tools

3.1 Voronoi Tesselation

Voronoi Tessellation and tiled regions are related concepts in spatial analysis. Voronoi Tessellation refers to a way of dividing a plane into regions based on the distance between a set of seed points, while tiled regions refer to the resulting partitions created by the Voronoi Tessellation.

Voronoi Tessellation is a mathematical concept that was first introduced by the Ukrainian mathematician Georgy Voronoi in 1908. Given a set of seed points in a plane, the Voronoi Tessellation partitions the plane into polygonal cells, with each cell being associated with a unique seed point. The cell of a seed point consists of all points in the plane that are closer to that seed point than to any other seed point. In other words, the Voronoi Tessellation creates a mapping between each seed point and the region of the plane closest to it.

The resulting tiled regions of the Voronoi Tessellation are often referred to as Voronoi diagrams [10]. These diagrams are used in a variety of fields, including computer science, geography, and engineering. For example, they can be used to model the distribution of resources, such as water or food, in a landscape, or to understand the patterns of communication between individuals in a social network.

Thus, Voronoi Tessellation and tiled regions are powerful tools for understanding the spatial relationships between points in a plane. By dividing a plane into regions based on distance, it is possible to gain insights into complex systems and make informed decisions based on the resulting information [11].

Mathematically, Voronoi tessellation can be defined as follows:

Let X be a set of n points in a d-dimensional space. For each point xi in X, the Voronoi cell V(xi) is defined as the set of all points in the space that are closer to xi than to any other point in X:

$$V(xi) = \{y \text{ in d-dimensional space} \mid dist(y, xi) <= dist(y, xi), \text{ for all } j != i\}$$

where:

d = the dimension of the space
X = the set of n input points
xi = a point in X
xi = another point in X, where j ! = i
dist(y, xi) = the Euclidean distance between the points y and xi

The Voronoi tessellation can be visualized as a partition of the space into cells, with each cell representing the Voronoi cell of a single point in X.

In summary, Voronoi tessellation is a mathematical concept that partitions a space into regions based on the proximity of points in the space. The Voronoi cell of a point xi is defined as the set of all points in the space that are closer to xi than to any other point in X.

3.2 Data Envelopment Analysis (DEA)

Data envelopment analysis (DEA) is a non-parametric programming method that has been widely used in the retail sector for measuring the efficiency of decision-making units (DMUs). It provides a framework for comparing the performance of multiple DMUs based on a set of inputs and outputs and enables decision-makers to identify best practices and areas for improvement. DEA has been applied in various aspects of retail operations, including store management, supply chain management, and customer service, among others.

The point of departure for the calculation of efficiency measures is the piece-wise linear frontier technology expressed by the following production possibility set:

where x is the input vector and y is the output vector, and in the last expression we have introduced J observations and indexed output by m and input by n. The variables λj ($j = 1 \dots J$) are non-negative weights (intensity variables) defining frontier points. Constant returns to scale is assumed for simplicity. The nature of scale does not matter for the question of model specification type with categorical variables [4].

3.3 Technology Stack

3.3.a. Storage. Google Drive is used for cloud storage and data access.

3.3.b. Languages. Python v3.8 via Google Colaboratory.

3.3.c. Libraries

- **geopandas:** Required for working with geospatial data.
- **contextily:** Used to retrieve and write to disk tile maps from the internet into geospatial raster files.
- **osmnx:** Used for downloading geospatial data from OpenStreetMap [1].
- **pystoned:** pyStoNED's DEA module is used to generate the technical efficiencies for a given set of inputs and outputs.

4 Implementation

4.1 Voronoi Tesselation

4.1.a Census Data

Following is the publicly available census data (Fig. 1) based on zipcodes of United

element_type	osmid	Store_ID	addr:city	addr:state	geometry	latitude	longitude
way	646640642	store-1	Hillsboro	OH	POINT (-83.62183 39.22779)	-83.621827	39.227794
	601084900	store-2	Newark	OH	POINT (-82.43068 40.08574)	-82.430679	40.085739
	461742392	store-3	Streetsboro	OH	POINT (-81.36204 41.25259)	-81.362036	41.252592
	86818680	store-4	Macedonia	OH	POINT (-81.51936 41.30998)	-81.519358	41.309975
node	8999419818	store-5	Aurora	OH	POINT (-81.38991 41.35005)	-81.389908	41.350055

Fig. 2. Stores and GIS information

States. Zipcodes of Ohio state were used to get the target population from tessellation output as explained further below.

4.1.b Store Location Data
Following table (Fig. 2) shows the store IDs and the GIS information for each of the stores which have been used for tessellation in the next steps.

4.1.c Store Location Map
Following (Fig. 3) is the map plot of the store locations in Ohio state.

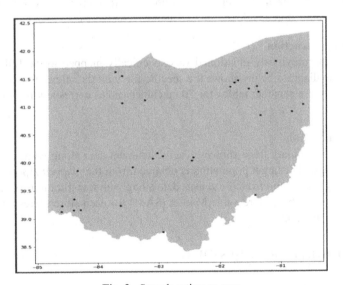

Fig. 3. Store location on map

4.1.d Tesselation Algorithm
Having obtained the state boundary and the coordinates for each store within the state, the data was fed into the tessellation algorithm for generating the catchment area i.e. tiled region for each store. Following is the Voronoi diagram as shown in Fig. 4. The

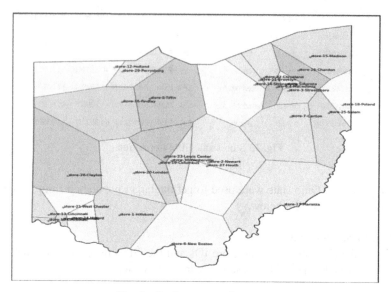

Fig. 4. Catchment area for stores

colored point within each tile represents a store, such that any point within each tile is nearest to the retail store inside the same (Euclidean distance).

4.1.e Sample Extraction
The census data previously mentioned was specifically mapped to the following tile map. The map diagram below shows the specific tile with the relevant zip codes on it. Similarly, there are separate tables for 30 catchment areas corresponding to 30 stores (Fig. 5).

4.1.f Summary Table
Following is a summary table showing each catchment area along with its respective target population. The target population is obtained from the mapped zipcodes and the corresponding population from the census data along with few filtrations on age group. This is the Serviceable Addressable Market (SAM) for each store corresponding to a specific catchment area.

4.2 Data Envelopment Analysis (DEA)

4.2.a DEA Asset Data
The catchment area population data, specifically, serviceable addressable market (SAM) as obtained after running the tessellation algorithm was combined with the operations' data for each of the 30 DMUs to create a composite dataset containing both input and output attributes for efficiency evaluation as next steps.

The above table in Fig. 7 shows a snapshot of 30 observations (5 observations shown) having all the dimensions of a store's name (DMU_Name), serviceable addressable market (SAM) on a 1:100000 scale, SAM density per sq. mile (Density) on a 1:10000 scale,

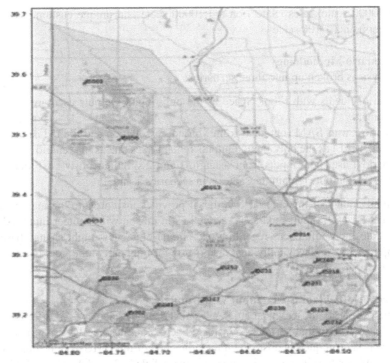

Fig. 5. Tile map with zipcodes

Tile_ID	DMU_Name	Population	Density	Tile_Region
0	store-1-Hillsboro	343958.0	10241.4	geometry POLYGON ((-84.7071902010505 39.639...
1	store-2-Newark	234584.0	21763.7	geometry POLYGON ((-84.50327486029713 39.17...
5	store-3-Streetsboro	1051588.0	41799.0	geometry POLYGON ((-83.91364314689105 40.42...
7	store-4-Macedonia	375534.0	11163.1	geometry POLYGON ((-83.8212025931618 41.304...
4	store-5-Aurora	504928.0	10129.4	geometry POLYGON ((-84.7071902010505 39.639...

Fig. 6. Summary table (5 obs. shown)

Tile_ID	DMU_Name	TAM	Density	Tile_Region	Floor_Space	Sales	Customers
0	store-1-Hillsboro	3.43958	1.02414	geometry POLYGON ((-84.7071902010505 39.639...	14	7	11.0
1	store-2-Newark	2.34584	2.17637	geometry POLYGON ((-84.50327486029713 39.17...	9	9	11.0
5	store-3-Streetsboro	10.51588	4.17990	geometry POLYGON ((-83.91364314689105 40.42...	12	7	8.0
7	store-4-Macedonia	3.75534	1.11631	geometry POLYGON ((-83.8212025931618 41.304...	12	20	20.0
4	store-5-Aurora	5.04928	1.01294	geometry POLYGON ((-84.7071902010505 39.639...	16	20	17.0

Fig. 7. DEA asset data

geometry data of the tiled region (Tiled_Region), floor space in sq. ft. (Floor_Space) on a 1:10000 scale, sales (Sales) on a 1:10000 scale and unique customers per day (Customers) on a 1:100 scale.

4.2.b Scenario Methodology
The analysis is broken up into three scenarios:

- Scenario 1 deals with Floor_Space as the only input. Outputs used are Sales and Customers.
- Scenario 2 uses SAM (obtained from Voronoi tessellation and filtered as discussed in Sect. 4.2.d.) as the only input. Outputs remain the same as in scenario 1.
- Scenario 3 uses both inputs from scenario 1 and 2 simultaneously i.e. Floor_Space and SAM. Outputs remain the same as in scenarios 1 and 2.

The advantage of this approach is that it provides insights into the maximization of outputs for each input, separately, as well as in unison with another input. Also, it demonstrates how inferences regarding operational efficiency might change based on each input metric, and why individual factors provide only part of the story in a practical business scenario.

4.2.c Scenario 1: Floor Space Input vs. Outputs
In the first scenario, the technical efficiency [2] is measured across DMUs for the given Floor_Space. The output variables considered here are Sales per day (value) and Customers per day (unique) against which the efficiency of the DMUs can be determined. Once the efficiency values are determined via the radial CCR model, the frontier is plotted accordingly. New derived output metrics are computed – Customers/FS and Sales/FS by dividing Customers and Sales respectively by Floor Space (FS) (Fig. 8).

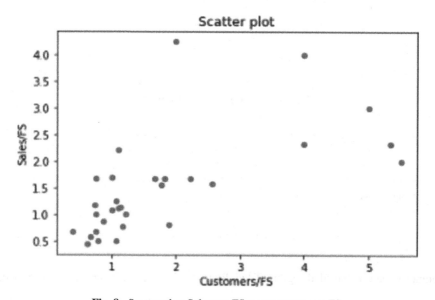

Fig. 8. Scatter plot: Sales per FS vs customers per FS

In the above scatter plot, both the axes are measuring two different kinds of outputs (Sales in Y axis and Customers in X axis), per unit of one input (Floor Space). Hence the X axis represents number of customers per unit floor space, while the Y axis represents number of sales per unit of floor space. It would always be preferable that X and Y values be maximum per unit of specific input measure taken into consideration for the above graph. Hence, if more than two points are connected, with those having the highest values in combination of X and Y measures, it will form a curve which is concave (with respect to the origin) in nature (it is an exception if more than 2 points comes in a straight line).

Hence, for maximizing outputs versus inputs, it would always be preferable that the curve which represents maximum efficiency be concave (with respect to origin) in nature. This curve is called as "frontier curve" for output-oriented analysis considering two outcomes versus one input measure [3]. A convex curve (with respect to the origin) will not be preferred as it will have points which are less efficient than desired ideally. Following the radial CCR output-oriented DEA model, efficiencies were computed for each of the 30 stores. The table below is a summary table for customers per floor space, sales per floor space and the efficiency computed previously (Table 1).

Table 1. Summary table for store efficiency with FS as input (5 obs. shown).

Index	Store	Customers/FS	Sales/FS	Efficiency
0	store-1-Hillsboro	0.785714	0.500000	0.160714
1	store-2-Newark	1.222222	1.000000	0.277778
2	store-3-Streetsboro	0.666667	0.583333	0.156250
3	store-4-Macedonia	1.666667	1.666667	0.416667
4	store-5-Aurora	1.062500	1.250000	0.307292

Table 2. Efficient DMUs with FS as input

Index	Store	Customers/FS	Sales/FS	Efficiency
14	store-15-Madison	2.000000	4.250000	1.0
9	store-10-Cincinnati	4.000000	4.000000	1.0
5	store-6-New Boston	5.000000	3.000000	1.0
19	store-20-London	5.333333	2.333333	1.0
23	store-24-Milford	5.500000	2.000000	1.0

The above table lists the efficient DMUs having efficiency equal to 1. The below chart shows the frontier curve plotted by joining the data points representing the efficient DMUs from the above tables in Scenario 1. Considering, that efficiency in this case relates to output maximization, the frontier curve is concave (with respect to origin) in nature as discussed previously (Fig. 9).

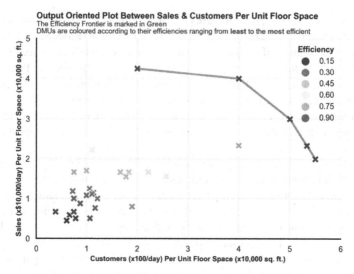

Fig. 9. Frontier curve for sales and customers with FS as input

4.2.d Scenario 2: SAM Input vs. Outputs

The second scenario is very similar to the first one where the technical efficiency across DMUs for the given SAM is measured. The output variables considered here are Sales and Customers as before, against which the efficiency of the DMUs can be determined. New derived output metrics are computed – Customers/SAM and Sales/SAM by dividing Customers and Sales respectively by SAM. Once the efficiency values are determined via the radial CCR model, the frontier is plotted accordingly (Fig. 11).

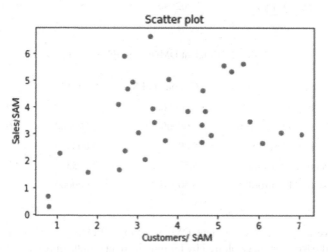

Fig. 10. Sales per SAM vs customers per SAM

As explained earlier in Scenario 1, both axes in the above scatter plot (Fig. 10) measure outputs per unit of a common input, in this case SAM. Hence, as mentioned before, if more than two points with the higher combination of values for X and Y axis are connected, the curve obtained would again be concave (with respect to origin) in nature. This would be the "frontier efficiency" curve.

It is to be noted that since outputs are being maximized, it would be expected that the frontier curve would preferably be concave (with respect to origin) in nature. However, if inputs were being minimized against fixed output to achieve better efficiency, the frontier curve would have preferably been convex (with respect to origin) in nature.

Following the radial CCR output-oriented DEA model, efficiencies were computed for each of the 30 stores. 3 stores have highest efficiency of 1.0 and others have opportunities for efficiency improvement. The below table lists the efficient DMUs i.e. DMUs having efficiency equal to 1.

Table 3. Efficient DMUs with SAM as input

Index	Store	Customers/FS	Sales/FS	Efficiency
15	store-16-Findlay	3.319447	6.638894	1.0
24	store-25-Salem	5.616564	5.618564	1.0
22	store-23-Lewis Center	7.068111	2.976047	1.0

The above plot (Fig. 11) shows the frontier curve plotted by joining the data points representing the efficient DMUs from the above tables in Scenario 2. Considering that efficiency in this case relates to output maximization, the frontier curve is concave (with respect to origin) in nature as discussed previously.

4.2.e Scenario 3: All Inputs and Outputs
For the third scenario, both inputs, i.e. Floor_Space and SAM, and both outputs, i.e. Sales and Customers are considered. These were fed into the radial CCR model as before to obtain the efficiencies.

Following the radial CCR output-oriented DEA model, efficiencies were again computed for each of the 30 stores. 8 stores have highest efficiency of 1.0 and other 22 have opportunities for efficiency improvement (Table 4).

Now, the distinguishing factor here is that a truly composite index [7] has been considered for efficiency comparison, which includes all prioritized inputs and outputs together. Though the holistic efficiencies of the individual stores are obtained in this scenario, it is not possible to generate meaningful graphs for visual interpretation as the dimensions are beyond 3.

4.3 Other Analyses

- Combo chart - Sales and number of employees vs. relative efficiencies across stores (Fig. 12):

- Combo chart - Sales and number of employees vs. relative efficiencies across stores (Fig. 13):

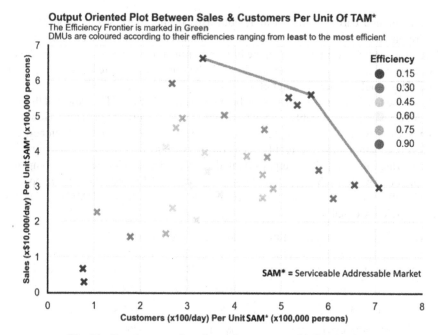

Fig. 11. Frontier curve for sales and customers with SAM as input

Table 4. Efficient DMUs with all inputs

Index	Store	Efficiency	SAM	Floor_space	Sales	Customers
3	store-4-Macedonia	1.0	3.75534	12	20	20.0
5	store-6-NewBoston	1.0	2.59080	3	9	15.0
9	store-10-Cincinnati	1.0	2.33751	2	8	8.0
14	store-15-Madison	1.0	7.52342	4	17	8.0
15	store-16-Findlay	1.0	3.01255	9	20	10.0
19	store-20-London	1.0	2.62282	3	7	16.0
22	store-23-LewisCenter	1.0	2.68813	10	8	19.0
24	store-25-Salem	1.0	1.24587	8	7	7.0

5   Outcomes

5.1   Findings: Tessellation and SAM

Serviceable addressable market (SAM) is important for retail store performance evaluation because it helps the store understand the potential size of the market and the level of demand for their products or services. By knowing the SAM, the store can estimate

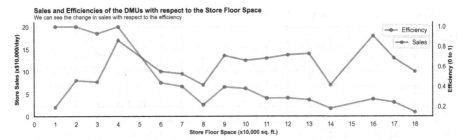

Fig. 12. Sales and efficiency variation against floor space

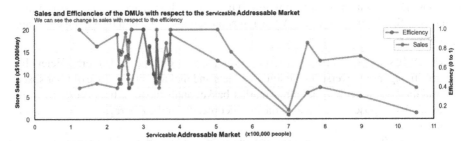

Fig. 13. Sales and efficiency variation against serviceable addressable market

their market share and identify opportunities for growth. For example, if the store is only capturing a small portion of the SAM, there may be opportunities to increase sales by expanding product offerings or targeting new customer segments.

However, estimating SAM may prove challenging for the following reasons:

- Difficulty defining the market: The definition of the market can be vague or subjective, making it challenging to determine the SAM. For a generic multi-brand retail chain, age group consideration is definitely critical.
- Market dynamics: The market can be constantly changing, with new competitors and shifting customer preferences, making it difficult to accurately estimate the SAM.
- Limited resources: Estimating the SAM can require significant time and resources, which may be limited for smaller businesses.

Keeping this in mind, probable SAM was computed from publicly available census data by using Voronoi tessellation to estimate the catchment areas of each store to provide geographic estimate of SAM population for each store after applying age filters.

The aim was to understand store performance based on the above SAM population when combined with regular operational factors such as store floor space, in order to get a holistic picture of store efficiency and address slacks [5].

The tessellation implementation given in Sect. 4.2.d used the available census data in Sect. 4.2. "Population" column (refer Fig. 6) represents the estimated SAM, taking into consideration the age profile of an average shopper and the catchment area obtained (indicated by "Tile Region") through tessellation. This information can then be used as a performance input indicator for efficiency evaluation across business units, in this case retail stores, using techniques such as Data Envelopment Analysis.

5.2   Findings: Single Input and 2 Outputs

The output metrics, namely Customers (unique) and Sales (value), were taken into account while determining efficiency values. The efficiency values were calculated per unit of the input metric - firstly, Floor_Space and secondly, serviceable addressable market or SAM.

Efficient DMUs i.e. DMUs with efficiency 1, were determined as per Sects. 4.2.c and 4.2.d respectively. 5 DMUs were identified on the efficiency frontier when Floor_Space was used as input metric. However, the number of efficient DMUs drops to 3 when SAM was used as input metric. (refer Tables 2 and 3).

Also, there were no common DMUs between the two efficient groups. This indicates that those DMUs which are performing well in terms of floor space are underperforming in terms of SAM, and vice versa. A few remedial actions that could be implemented are as follows:

- The DMUs underperforming in terms of floor space could be made more efficient by prioritizing more diverse and unique products and making efficient use of floor space available via adjacency planning (market basket analysis etc.). Improving the store layout that exposes customers to the product breadth and encourages exploration can also help.
- The DMUs lacking in terms of serviceable addressable market could improve through better customer outreach programs, targeted advertising and promotional campaigns through specific customer loyalty programs based on individual buying patterns. Expanding the retail store's availability through hybrid mode (e-commerce door delivery or pre-order pick-up facility) can help reach a larger conversion from the available SAM.
- Better inventory optimization along with understanding of substitute products and customer feedback analysis would help stock availability as well as increased customer satisfaction. Implementing such changes would increase the output for the underperforming DMUs, thereby pushing them towards higher efficiency.

5.3   Findings: All Inputs and Outputs

5.3.a. Findings 1. Following are DMUs with different inputs, comparable efficiency (Table 5).

Table 5. DMUs with different inputs but comparable efficiency

Index	Store	Efficiency	SAM	Floor_Space	Sales	Customers
10	store-11-Brooklyn	0.477595	3.35366	12	8	9.0
11	store-12-Holland	0.429580	7.86889	17	13	20.0

5.3.b Findings 2. - Following are DMUs with lower outputs but higher efficiency (Table 6).

Table 6. DMUs with lower outputs but higher efficiency

Index	Store	Efficiency	SAM	Floor_Space	Sales	Customers
24	store-25-Salem	1.000000	1.24587	8	7	7.0
25	store-26-SouthEuclid	0.891646	3.37864	12	20	9.0

5.3.c Findings 3. - Following are DMUs with similar inputs but different outputs and efficiencies (Table 7).

Table 7. DMUs with similar inputs but different outputs and efficiencies

Index	Store	Efficiency	SAM	Floor_Space	Sales	Customers
5	store-6-New Boston	1.000000	2.59080	3	9	15.0
16	store-17-Marietta	0.794872	2.61031	3	7	12.0

5.3.d Findings 4. - Following are DMUs which are best performing stores (Table 8).

Table 8. Best performing DMUs

Index	Store	Efficiency	SAM	Floor_Space	Sales	Customers
3	store-4-Macedonia	1.0	3.75534	12	20	20.0
5	store-6-New Boston	1.0	2.59080	3	9	15.0
9	store-10-Cincinnati	1.0	2.33751	2	8	8.0
14	store-15-Madison	1.0	7.52342	4	17	8.0
15	store-16-Findlay	1.0	3.01255	9	20	10.0
19	store-20-London	1.0	2.62282	3	7	16.0
22	store-23LewisCenter	1.0	2.68813	10	8	19.0
24	store-25-Salem	1.0	1.24587	8	7	7.0

5.4 Findings: Charts

5.4.a Findings. Floor space as input (Refer Sect. 4.3 Fig. 12).

- Sales have a general upward trend with increasing floor space but with periodic spikes and dips. This is expected behavior since increase in SAM would also increase the number of casual window shoppers as well.

- Stores with less floor space and lower sales show the highest efficiency because of better space utilization.
- Average efficiency is high with lower floor space but as the floor space increases efficiency gains a downward trend due to poorer space utilization. Adjacency analysis could prove useful in this case.

5.4.b Findings: SAM as input (Refer Sect. 4.3 Fig. 13).

- Sales are generally increasing with the increase in serviceable addressable market; however, there are some sudden sharp dips. This could likely be attributed to factors such as insufficient inventory, understaffing, or external global crises such as the coronavirus pandemic. Frequent promotional offers and discounts can counter this to a sizeable extent.
- Stores with a smaller SAM as well as fewer sales, have higher efficiency. These stores would likely fall under the "boutique" category with a dedicated group of customers and serving a smaller area.

6   Discussion

Under DEA, there are methods for both input minimization and output maximization. In retail industry, investments in real estate and infrastructure are already made prior while setting up stores. Also, SAM or serviceable market in a geography or investments/commitments in supply chain does not change often in near term horizon. So, in retail businesses, there is limited opportunity in minimizing input parameters though businesses want to maximize the output metrics to improve efficiency. Hence output maximization DEA variant is appropriate as has been used in this case study for the retail chain to compare and benchmark performance of various stores. Also, the output-oriented DEA method uses constant ratio to scale (CRS) variant which is effective but simple method for computing outputs per unit of inputs for defining the frontier envelope as showcased in the implementation above.

6.1 Increase in Number of Efficient Units - Based on the findings and recommendations discussed and implemented in Sect. 5, the number of efficient DMU's which were 8 stores in Sect. 5.3.d (based on 2 inputs and 2 outputs) increased further to 13 stores.

6.2 Composite Index for Multiple Inputs and Outputs - As demonstrated via the experiments conducted in this case, it is necessary to form a composite index of all operational performance metrics for a holistic view of unit performance. This enables in analysis of individual efficiency measures along with composite measures which leads to better business decisions by managers.

6.3 Some External Factors Cannot Be Directly Manipulated - However, it is to be noted that some factors like floor area and total target addressable market (SAM) may not be as easily optimized as other factors. Therefore, the opportunity lies in maximizing total output for the fixed input metrics. Hence, it is important to decide which model is more relevant - maximizing outputs or minimizing inputs to compare and benchmark efficiency.

6.4 Use of Output Maximization Model for Fixed Inputs - As demonstrated in this paper, operational efficiency based on output maximization via Data Envelopment Analysis (DEA) following identification of key input metrics like serviceable addressable market obtained via tessellation using census and store data can lead to various important insights related to target population.

6.5 Visual Inferences - Visual insights obtained from plotting sales and efficiency with respect to floor space and serviceable addressable market respectively lead to interesting data trends pertaining to the changes in efficiency and sales based on the input metrics. This provides better explainability by supplementing the results from DEA with actionable items for further consideration and analysis.

7 Conclusion

This case study used a combination of algorithms and certain available dimensions to enable business decision-makers to identify efficient stores and drive improvements in the stores. Dimension priority methods can also be deployed using either regression, principal component methods (PCM) or analytical hierarchical process (AHP) as any business will have many inputs and outputs measures and it is important to consider vital few instead of trivial many. Buyers of any retail outlet can live in a semi-urban locality and commute to office in an urban locality and visit rural areas within a state very frequently. So, SAM determination is not easy and never precise. Fuzzy set theory can be applied to DEA models to handle imprecise inputs [6] for considering these situations by defining tolerance levels on the objective function and constraint violations. Also, competitor information plays a vital role in decision making while comparing and benchmarking retail store efficiency. Public data (with available competitor information on stores) can be accessed more comprehensively to account for the same to further the robustness of the analysis outcomes. This approach is also applicable to other verticals of healthcare [12, 13] and education [14, 15] to compare performance of healthcare units/clinics or education centers (schools, colleges etc.) as decision making units (DMUs) with relevant input and output parameters.

It should also be noted that the various studies and work done with DEA and tessellation, independent of each other, as publicly available are from different countries and demographic regions. So, any constraint, applicable to the framework as discussed above, while extending the same to other specific geographic regions is not known.

References

1. Boeing, G.: OSMnx: new methods for acquiring, constructing, analyzing, and visualizing complex street networks. Comput. Environ. Urban Syst. **65**, 126–139 (2017)
2. Charnes, A., Cooper, W.W., Lewin, A.Y., Seiford, L.M.: Data envelopment analysis: theory, methodology, and applications, 23–47 (1995)
3. Pascoe, S., Kirkley, J.E., Gréboval, D., Morrison-Paul, C.J.: Measuring and assessing capacity in fisheries 2. Issues and methods, appendix D (2003)
4. Førsund, F.R.: Categorical variables in DEA. Int. J. Bus. Econ. **1**(1), 33–43 (2002)

5. Malik, M., Efendi, S., Zarlis, M.: Data envelopment analysis (DEA) model in operation management. IOP Conf. Ser. Mater. Sci. Eng. **300**, 012008, 3–4 (2018)
6. Valami, H.B., Nojehdehi, R.R., Abianeh, P.M.M., Zaeri, H.: Production possibility of production plans in DEA with imprecise input and output. Res. J. Appl. Sci. Eng. Technol. **5**(17), 4264–4267 (2013)
7. Wua, H., Yang, J., Chen, Y., Liang, L., Chena, Y.: DEA-based production planning considering production stability. INFOR Inf. Syst. Oper. Res. **57**(3), 477–494 (2019)
8. Lawrence, K.D., Kudyba, S., Klimberg, R.K., Lawrence, S.M.: A DEA efficiency analysis of operating units within electronic shopping stores. Appl. Manag. Sci., 1–2 (2013)
9. Donthu, N., Yoo, B.: Retail productivity assessment using data envelopment analysis. J. Retail. **74**(1), 89–105 (1998). ISSN: 0022-4359
10. Aurenhammer, F.: Voronoi diagrams – a survey of a fundamental geometric data structure. ACM Comput. Surv. **23**(3), 345–405 (1991)
11. Sen, Z.: Spatial Modeling Principles in Earth Sciences. Springer, Cham, p. 57 (2016). https://doi.org/10.1007/978-3-319-41758-5
12. Babaqi, T., Vizvári, B.: The post-disaster transportation of injured people when hospitals have districts. J. Humanit. Logist. Supply Chain Manag. (2023)
13. Zakowska, I., Godycki-Cwirko, M.: Data envelopment analysis applications in primary health care: a systematic review. Fam. Pract. **37**(2), 147–153 (2019)
14. Masouleh, F.A.N., Murayama, Y., Rho'Dess, T.W.: The application of GIS in education administration: protecting students from hazardous roads. Trans. GIS **13**(1), 105–123 (2009)
15. Johnes, J.: Data envelopment analysis and its application to the measurement of efficiency in higher education. Econ. Educ. Rev. **25**(3), 273–288 (2006)
16. Kurek, K.A., Heijman, W., van Ophem, J., Gędek, S., Strojny, J.: Measuring local competitiveness: comparing and integrating two methods PCA and AHP. Qual. Quant. **56**(3), 1371–1389 (2022)

A Machine Learning Framework Developed by Leveraging the Topological Pattern Analysis of Bitcoin Addresses in Blockchain Network for Ransomware Identification

Dipayan Chakraborty[1]([✉]) [iD], Sangita Mazumder[2] [iD], and Ashutosh Kar[2] [iD]

[1] TATA Consultancy Services, Kolkata, India
dipayanchakrabortymail@gmail.com
[2] NSHM Knowledge Campus Kolkata, Kolkata, India

Abstract. The steep uprise of the incognito cryptocurrency transactions has significantly raised the chances for the ransomware developers of demanding ransom by enciphering sensitive data of not only the individuals but also the large corporate houses.

Majority of the recent ransomware operators prefer bitcoin to be a medium of their murky transactions. Although these bitcoin postings are permanently documented in the blockchain ledger, existing practices to detect & mine the origin of these ransomwares are still very difficult, which consist of numerous data gathering steps – making the process much lengthy & complex.

As a solution to that, multiple statistical tools machine learning algorithms like Decision Tree, Random Forest, Binomial & Multinomial Logistic Regression has been utilized and a framework has been suggested to detect malicious addresses automatically with a considerably low lead time, by capturing the input data from coin movement topology while leveraging the graph theory & network analysis methodologies as its foundation.

Keywords: Bitcoin · Machine Learning · Decision Tree · Random Forest · Logistic Regression

1 Introduction

For more than a decade, the blockchain based technologies have seen a drastic uprise. Basically, blockchain is such a distributed public ledger that stores transactions between a couple of parties, where a bona fide centralized authority is not required at all. On a blockchain network, two unaccounted parties can orchestrate a rigid transaction that is forever listed down on the blockchain ledger, publicly. One of the very beginning implications of Blockchain has been the Bitcoin cryptocurrency. Presently there are more than thousand Blockchain based cryptocurrencies existing in the macroenvironment ecosystem.

The extensive and steep rise of cryptocurrencies (e.g., Bitcoin in the current study) that allow pseudo-anonymous transactions has made it easier for ransomware developers to demand ransom by encrypting sensitive user data. The recently revealed strikes of ransomware attacks have already resulted in significant economic losses and societal harm across different sectors, ranging from common people to big corporates.

S. Dhar et al. (Eds.): AGC 2023, CCIS 2008, pp. 23–39, 2024.
https://doi.org/10.1007/978-3-031-50815-8_2

Most modern ransomware use Bitcoin for payments. However, although the Bitcoin postings are tracked in such a way which has no expiry date (lasting forever) and publicly available, current way-arounds to arrest the ransomware are dependent on only a handful techniques and/or lengthy information collection steps.

By capitalizing on the recent advances in topological data analysis, an effective and controllable machine learning framework has been formulated through the current research in order to automatically detect new malicious addresses in a particular ransomware family, given only a limited record of previous transactions.

2 Objectives of the Study

In this study, the objectives are:

- To understand and analyse various topological patterns of Bitcoin addresses in the blockchain network.
- To design a data-driven Bitcoin transaction analytics framework which is based upon historical data & is significantly effective in detecting ransomware payment related addresses,
- To ensure that the aforesaid data analytics framework is featured to classify the ransomware payment related addresses apart from the valid and legal transactions of Bitcoin payments.
- To apply the algorithm and successfully identify (as far as possible) the Bitcoin addresses that are used to store and trade Bitcoins gained through ransomware activities.

3 Literature Review

Satoshi Nakamoto (2008) has proposed a solution which can propose an offering to the double-spending problem using a peer-to-peer network in online payments. Network timestamps transactions needs to be hashed into a seamless chain of hash-based proof-of-work, which could be formed in such a way that cannot be changed by not redoing the proof-of-work. Also, the longest chain will serve as proof of the sequence of events witnessed and the network also requires minimum structure. In this dedicated ecosystem, nodes can leave and re-join the network at their will, accepting the longest proof-of-work chain as a proof or exhibit of what happened while they were gone. E. Androulaki, G. O. Karame, M. Roeschlin, T. Scherer, and S. Capkun (2013) have investigated the root cause of the phenomena; that although Bitcoin is quickly emerging as a popular digital payment system throughout the globe; raises lot of concerns related with the truth that all of these transactions/postings that are taking place are publicly declared in the given system. The study has exhibited that the profiles of almost 40% of the users can be, to a large extent, recovered even when users adopt privacy measures recommended by Bitcoin. Also, M. Ober, S. Katzenbeisser, and K. Hamacher (2013) have investigated the effect of network structure & network dynamics on the anonymity challenges of Bitcoin payment systems. G. Maxwell (2013) has illustrated that a poor privacy in Bitcoin can be a major practical disadvantage for both individuals and businesses; leading to create externalized costs & negative credibility of the entire payment ecosystem. S. Meiklejohn, M. Pomarole, G. Jordan, K. Levchenko, D. McCoy, G. M. Voelker, and S. Savage (2013) have explored the unique characteristic of visible flow with anonymous ownership featured by the Bitcoin transactions, using the heuristic clustering framework in order to group the Bitcoin wallets based on evidence of shared authority, and then using re-identification influxes or the empirical purchasing of goods and services, to classify the operators/administrators of those clusters. T. Ruffing, P. Moreno-Sanchez and A. Kate (2014) has proposed an entirely decentralized Bitcoin mixing protocol named CoinShuffle; which allows its users to utilize Bitcoin in an absolutely anonymous manner and been inspired by the accountable anonymous group communication protocol Dissent it enjoys numerous advantages over its predecessor Bitcoin mixing protocols including total avoidance of any additional anonymization charges and computation & communication overhead minimization for the whole framework. N. Andronio, S. Zanero, and F. Maggi (2015) has described the common features of various ransomware families and featured a framework, to detect the patterns if an app is attempting to encrypt or lock the device without the consent of the user i.e., an approach that differentiates known and unknown ransomware and malware samples from non-ransomware & non-malware samples. G. Di Battista, V. Di Donato, M. Patrignani, M. Pizzonia, V. Roselli, and R. Tamassia (2015) has described a system for the visual analysis of how and when a flow of Bitcoins mixes with other flows in the transaction graph which also focuses on how a Bitcoin transaction system depends on high-level metaphors the depiction of the graph and the size and characteristics of the transactions, further allowing for high level analysis of big portions of it. F. Tschorsch and B. Scheuermann (2016) has shown how exploiting the Bitcoin protocol and its building blocks, fundamental structures and insights at the core of the Bitcoin protocols could be deduced and utilized in various analytical frameworks. K. Liao, Z. Zhao, A. Doupe, and G.-J. Ahn (2016) conducted a measurement analysis of a ransomware family called CryptoLocker that encrypts a victim's files until a ransom is been paid, within the Bitcoin ecosystem between the timestamp from September 5, 2013 to January 31, 2014 & identified 795 ransom payments totaling 1,128.40 BTC ($310,472.38) by constructing &analyzing a network topology for obtaining auxiliary information regarding the operation of the Cryptolocker ransomware family. This study was also engaged in underscoring the measurement analysis value and provided threat intelligence in understanding the erratic cyber-crime landscape, by featuring a prototype. A Narayanan and M. Moser (2016) have elucidated sixteen privacy-preserving techniques for Bitcoin on an axis of obfuscation-vs.-cryptography and highlighted the practice of obfuscation in order to balance between operator privacy and overall regulatory acceptance. M. Paquet-Clouston, B. Haslhofer, and B. Dupont (2018) have presented a light weight framework using a holistic view of its genesis, development, the process of infection and execution, and characteristic of ransom demands for each and every probed ransomware; to identify, collect, and analyze Bitcoin addresses which are basically managed by the same user or group of users which includes a novelty driven pattern for classifying a payment as ransom or ransomware associated transaction. D. Y. Huang, D. McCoy, M. M. Aliapoulios, V. G. Li, L. Invernizzi, E. Bursztein, K. McRoberts, J. Levin, K. Levchenko, and A. C. Snoeren (2018) created a measurement framework for ransomware identification. To build the experiment ecosystem, an array of data sources, including ransomware binaries, seed ransom payments, victim telemetry from infections, and a large database of Bitcoin addresses annotated with their owners were combined together in order to drive a large-scale, two-year, end-to-end measurement of ransomware payments, victims, and operators. i.e., the described statistical framework actually identified shared hacker behavior and used heuristics to identify the ransomware payments. M. Conti, A. Gangwal, and S. Ruj (2018) exploited the 360° view of the genesis, development, the process of infection and execution, and characteristic of ransom demands in order to formulate a lightweight statistical framework to identify, collect, and analyze Bitcoin addresses managed by the same user or group of users having a novel pattern of transaction characteristics; which have actually analyzed networks of crypto-currency ransomware, and identified that the attribute, traits & behavior of a hacker (s) could be useful to classify the ransomware payments that are still not disclosed. Martin, J. Hernandez-Castro, and D. Camacho (2018) have focused on highly complex & fast spreading Android malwares which hinders the use of reverse engineering techniques and anti-malware tools by deploying Advanced obfuscation, emulation detection, delayed payload activation or dynamic code loading; in a Bitcoin transaction environment. Also, Akcora, Cuneyt & Li, Yitao & Gel, Yulia & Kantarcioglu, Murat (2019) have capitalized on the latest advances in topological data analysis and came up with a robust and controllable data analytics framework to automatically detect new malicious addresses in a ransomware family.

Besides, Martín, A., Hernandez-Castro, J., Camacho, D (2018) has professed a detailed study on a specific family of android ransomwares called Jisut and prescribed a novel framework to tackle the said family of ransomware. Alongside, Rivera-Castro, R., Pilyugina, P., Burnaev, E. (2019) has applied Topological Data Analysis from the perspective of the topological structure of the Cryptocurrency datasets in order to enable an optimal portfolio management of cryptocurrencies. Besides, Goldsmith, D., Grauer, K., Shmalo, Y (2020) have established throgh their research that temporal features of a cryptocurrency topological network, are more useful than that of the static features like node balances & in and out degrees.

4 Research Methodology

Since the objective is to identify those Bitcoin addresses which are potential ransomwares and are associated with fraudulent transactions; a classification model (having multiple metric input and binary output) can be formulated in order to classify/label the Bitcoin addresses into whitelisted & blacklisted labels.

Furthermore, as a value addition component, it could be moved forward and build a statistical framework in order to classify the suspected Bitcoin addresses under various ransomware groups.

4.1 Statement of Problem

To identify Bitcoin addresses that are used to store and trade Bitcoins gained through ransomware & fraudulent activities by analyzing the patterns of the Bitcoin address topology in the Blockchain network.

4.2 Research Design

In this study, prior to formulate a design for the research, there are four basic queries that comes from a researcher's point of view while analyzing the patterns of Bitcoin addresses in the blockchain network. They are:

- What features extracted from the Bitcoin network can be used to detect ransomware behavior?
- Does a particular family of ransomwares show the recurrent traits on the Bitcoin blockchain over the time? Precisely, do they show a common pattern all along?
- How far known is the trait/attribute of various ransomware promoters/operators on the Bitcoin blockchain?
- Could it be possible to detect Bitcoin ransom payments using a data driven statistical model?

On the Bitcoin network, a given Bitcoin address can be present in more than one time with multiple inputs and outputs. In order to mine address behavior in time, the Bitcoin network could be divided into 24-h long windows.

This window approach practically serves two purposes:

First, the incited 24-h network of Bitcoin blockchain permits one to arrest how frequent a given coin can travel in the network. The speed is measured by the number of blocks in the 24-h window that contain a transaction involving the coin. In maximum, a coin can appear in 144 blocks (24 h, i.e. 6 blocks per hour). The velocity of a coin might give particular significant information on the transaction's purpose. For example, Ponzi schemes & various kinds of suspicious money laundering instances (i.e., investment fraud, tax evasion, drug/narcotics trafficking, smuggling of drug & narcotics, bribery-and-corruption and illegal/black money-racketeering) can be highlighted by using the greater pulsation identified in the movement of certain coins in the blockchain network.

Second, the time-based order of transactions within the window helps one distinguish the transaction activity from different geographical locations. Time-based (temporal) information of transactions, such as the local time, has also been found helpful to classify the criminal transactions.

In the next page there is a pictorial representation of the window approach of recording the Bitcoin transactions in the blockchain network (Fig. 1).

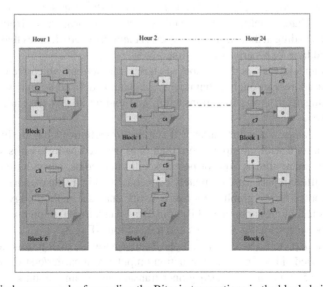

Fig. 1. Window approach of recording the Bitcoin transactions in the blockchain network.

Here, the coins c2 & c3 are those coins who have moved too frequent in the network than the others. It could be presumed that these coins are associated to the malpractices of money-laundering or other murky activities.

Hence the addresses related to these coins are needed to be analyzed using an appropriate data mining framework in order to find any sort of underlying fraudulent activities.

On the heterogeneous Bitcoin network, in each snapshot the following six features are been extracted for an address: income, neighbors, weight, length, count, and loop (Fig. 2).

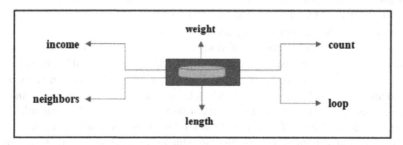

Fig. 2. Features of Bitcoin Addresses

Income of an address a is the total amount of coins output to that particular address a.

Neighbors of an address a is the number of transactions which have that particular address as one of its output addresses.

Weight of an address a is defined as the sum of fraction of coins that originate from a starter transaction and reach that very particular address a.

Length of an address a is the number of non-starter transactions on its longest chain, where a chain is defined as an acyclic directed path originating from any starter transaction and ending at that particular address. A length of zero denotes that the address is an output address of a starter transaction.

Count of an address a is the number of starter transactions which are connected to that particular address through a chain, where a chain is defined as an acyclic directed path originating from any starter transaction and ending at that particular address a mentioned above.

Loop of an address a is the number of starter transactions which are linked to that particular address a with multiple (more than one) directed path. Loop is supposed to count that how many transactions or postings split their coins, carry these coins in the network via different paths and ultimately converge them in a single address.

Largely rooted in the graph theory & network analysis methodology, the Bitcoin data analytics techniques approached Bitcoin data by creating a graph that employs the relationship between Bitcoin addresses and transactions (Fig. 3).

The figure depicted above, is a sample Bitcoin network of 8 transactions and 12 addresses. Dashed sides represent transaction outputs from earlier/erstwhile windows; tx1, tx3, tx4, tx5 & tx8 are starter/beginner transactions. Coin amounts are shown on

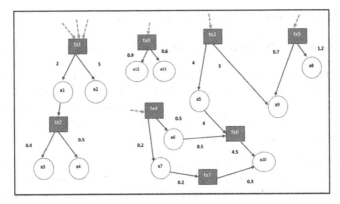

Fig. 3. Sample Bitcoin Network

edges. In case, Transaction outputs are equal to transaction inputs; i.e., transaction fees are zero (0).

As the part of the aforesaid framework, below classification models will be formulated in order to answer the problem statement:

- Decision Tree
- Random Forest (for classification)
- Binary Logistic Regression model

Also, another algorithm could be formulated: Multinomial Logistic Regression model in order to classify the Bitcoin addresses according to the ransomware groups/families.

4.3 Research Design

The dataset is at its best if it were randomized in row because it is undesirable to train a portion of the datasets which contains large proportion of loan default and using that to test on a dataset with small proportion, and vice versa. This will generate biasness for the machine learning algorithm rules and will result into undesirable outputs.

Therefore, all the observations were permutated and stored in an object. This object contains a vector of as many numbers of rows the dataset has, which are randomized out of the overall possibilities.

After that, the randomized dataset will be divided into a training sample to train the statistical model and a test sample to check the effectiveness of it.

Sources of Data Collection: In order to build the data driven Bitcoin transaction analytics framework, secondary data has been collected from the below machine learning database repository which is broadly used by machine learning community for the empirical analysis of various machine learning algorithms.

https://archive.ics.uci.edu/ml/datasets/BitcoinHeistRansomwareAddressDataset.

5 Data Analysis and Interpretation

5.1 Formulation of a Decision Tree Model for Classification

Since all the predictor parameters (i.e., income, neighbors, weight, length, count and loop) are continuous in nature, a Decision Tree could be easily formulated in order to classify all the addresses between two classes-Black Labeled (the addresses which are associated with ransomwares) & White Labeled (the addresses which are not associated with ransomwares).

Pre-processing of Data: Since in the current dataset, there are no any as such impurities or errors like "missing data", "null value", "NA" entries, the only pre-processing activity will be removing the biasness from the dataset by randomizing it and all the observations will be permutated and stored in an object. This object contains a vector of as many numbers of rows the dataset has, which are randomized out of the overall possibilities. After that, the randomized dataset will be divided into a training sample to train the statistical model and a test sample to check the effectiveness of it. 50% of the entire dataset will be used to train the model and the rest 50% data will be used to validate the model i.e., to check the effectiveness of it (Table 1).

Table 1. Decision Tree Output

```
Evaluation on training data (5117 cases):

           Decision Tree
           ----------------
     Size       Errors

       51   477( 9.3%)    <<


      (a)    (b)     <-classified as
      ----   ----
      828    299     (a): class Black
      178   3812     (b): class white

    Attribute usage:

    100.00% count
     95.11% income
     88.47% neighbors
     39.32% length
     35.51% weight
      1.58% looped
```

Evaluation with Training Data: Here true positive is real Black classified as black as our goal is to find ransomware addresses/blacklisted addresses and True negative is real white classified as whiteFalse positives are those addresses which are actually whitelisted but wrongly classified as blacklisted and false negatives are those addresses which are actually blacklisted but wrongly classified as whitelisted. The correctness in predictability here is 90.68%, (as per the confusion matrix) which denotes that the model is quite healthy in classifying the addresses (Table 2).

Table 2. Confusion matrix for the training data (decision tree)

```
dt_pred Black white
  Black   828    178
  white   299   3812
```

Evaluation with Test Data: Here true positive is real Black classified as black as our goal is to find ransomware addresses/blacklisted addresses and True negative is real white classified as white. False positives are those addresses which are actually whitelisted but wrongly classified as blacklisted and false negatives are those addresses which are actually blacklisted but wrongly classified as whitelisted. The correctness in predictability here is 66.52%, (as per the confusion matrix) which denotes that the model is quite healthy in classifying the addresses (Table 3).

Table 3. Confusion matrix for the test data (decision tree)

```
dt_pred Black white
  Black   192    814
  white   899   3212
```

5.2   Formulation of a Random Forest Model for Classification

Since all the predictor parameters (i.e., income, neighbors, weight, length, count, and loop) are continuous in nature, a Random Forest could be easily formulated in order to classify all the addresses between two classes - Black Labeled (the addresses which are associated with ransomwares) & White Labeled (the addresses which are not associated with ransomwares) (Table 4).

Table 4. Random Forest Output

```
Call:
 randomForest(formula = as.factor(label) ~ ., data = TrainSet,
importance = TRUE)
                  Type of random forest: classification
                          Number of trees: 500
No. of variables tried at each split: 2

        OOB estimate of  error rate: 9.6%
Confusion matrix:
       Black white class.error
Black   808   338  0.29493892
white   153  3818  0.03852934
```

Evaluation with Training Data: Here true positive is Black (as our goal is to find ransomware addresses/blacklisted addresses) and true negative is white (whitelisted addresses). False positives are those addresses which are actually whitelisted but wrongly classified as blacklisted and false negatives are those addresses which are actually blacklisted but wrongly classified as whitelisted. The correctness in predictability here is 95.43%, (as per the confusion matrix) which denotes that the model is quite healthy in classifying the addresses (Table 5).

Table 5. Confusion matrix for the training data (Random Forest)

```
predTrain Black white
    Black   989    77
    white   157  3894
```

Evaluation with Test Data: Here true positive is Black (as our goal is to find ransomware addresses/blacklisted addresses) and true negative is white (whitelisted addresses). False positives are those addresses which are actually whitelisted but wrongly classified as blacklisted and false negatives are those addresses which are actually blacklisted but wrongly classified as whitelisted. The correctness in predictability here is 89.86%, (as per the confusion matrix) which denotes that the model is quite healthy in classifying the addresses (Table 6).

Table 6. Confusion matrix for the test data (Random Forest)

```
predValid Black white
    Black   727   174
    white   345  3871
```

5.3   Formulation of a Binary Logistic Model for Classification

Referring to this, since all of the input variables i.e. income, neighbors, weight, length, count, loop are continuous in nature and the output variable i.e. the label (the classes) is a binary categorical variable - Black Labeled (the addresses which are associated with ransomwares) & White Labeled (the addresses which are not associated with ransomwares); a Binomial logit model or a binomial logistic regression model could be used in order to classify the Bitcoin addresses as per the above classification (Table 7).

Table 7. (a), (b), (c). Binary Logistic Regression Output

```
Call:
glm(formula = label ~ length + weight + count + looped + neighbors +
    income, family = binomial(link = "logit"), data = trainingData)

Deviance Residuals:
     Min       1Q   Median       3Q      Max
  -5.5644   0.0000   0.1726   0.5370   2.2292

Coefficients:
              Estimate Std. Error z value Pr(>|z|)
(Intercept)  1.836e-01  8.094e-02   2.268  0.02331 *
length       3.558e-02  2.990e-03  11.901  < 2e-16 ***
weight      -2.709e-01  9.673e-02  -2.800  0.00511 **
count       -1.112e-01  1.442e-02  -7.712 1.23e-14 ***
looped       3.835e-02  2.943e-02   1.303  0.19260
neighbors   -4.553e-02  3.760e-02  -1.211  0.22584
income       9.012e-10  4.825e-11  18.679  < 2e-16 ***
---
Signif. codes:  0 '***' 0.001 '**' 0.01 '*' 0.05 '.' 0.1 ' ' 1

(Dispersion parameter for binomial family taken to be 1)

    Null deviance: 5400.6  on 5116  degrees of freedom
Residual deviance: 3236.7  on 5110  degrees of freedom
AIC: 3250.7

Number of Fisher Scoring iterations: 13
```

(a)

```
$Models
Model: "glm, label ~ length + weight + count + looped + neighbors + income
, binomial(link = \"logit\"), trainingData"
Null:  "glm, label ~ 1, binomial(link = \"logit\"), trainingData"

$Pseudo.R.squared.for.model.vs.null
                         Pseudo.R.squared
McFadden                         0.400672
Cox and Snell (ML)               0.344842
Nagelkerke (Cragg and Uhler)     0.528935

$Likelihood.ratio.test
 Df.diff LogLik.diff  Chisq p.value
      -6     -1081.9 2163.9       0

$Number.of.observations
Model: 5117
Null:  5117

$Messages
[1] "Note: For models fit with REML, these statistics are based on refitti
ng with ML"

$Warnings
[1] "None"
```

(b)

```
   Hosmer and Lemeshow test (binary model)

data:  trainingData$label, fitted(logitMod)
X-squared = 18535, df = 8, p-value < 2.2e-16
```

(c)

Evaluation with Training Data: Misclassification error is 14.58% here, which means that there is 85.42% chance of correct classification using this training dataset by the model. Concordance is above 0.90 which denotes it as a good predictive algorithm. As far as the confusion matrix is concerned, blacklisted is denoted as 0 and whitelisted is denoted as 1 here. (Here true positive is real Black classified as black as our goal is to find ransomware addresses/blacklisted addresses and True negative is real white classified as white). Since both the sensitivity (true positive rate) and specificity (true negative rate) which are 0.89 & 0.70 accordingly and greater than the optimum cut-off of the probability value (0.58 for this model with training dataset), we can state that this is a good fit (Table 8).

Table 8. (a), (b), (c). Confusion Matrix, Concordance value & ROC curve for the training data (Binary Logistic Regression)

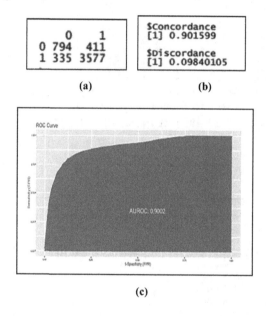

(a)   (b)

(c)

Evaluation with Test Data: Misclassification error is 14.68% here, which means that there is 85.32% chance of correct classification using this training dataset by the model. Concordance is above 0.89 which denotes it as a good predictive algorithm. As far as the confusion matrix is concerned, blacklisted is denoted as 0 and whitelisted is denoted as 1 here. (Here true positive is real Black classified as black as our goal is to find ransomware addresses/blacklisted addresses and True negative is real white classified as white). Since both the sensitivity (true positive rate) and specificity (true negative rate) which are 0.92 & 0.60 accordingly and greater than the optimum cut-off of the probability value (0.53 for this model with training dataset), we can state that this is a good fit (Table 9).

Table 9. **(a), (b), (c).** Confusion Matrix, Concordance value & ROC curve for the test data (Binary Logistic Regression)

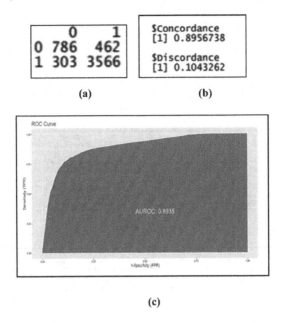

	0	1
0	786	462
1	303	3566

```
$Concordance
[1] 0.8956738

$Discordance
[1] 0.1043262
```

(a)   (b)

(c)

5.4 Formulation of a Multinomial Logistic Regression Model for Classification

Multinomial regression is used to forecast the nominal target variable. Since in this model the goal is to predict the Ransomware families the fraudulent Bitcoin addresses are associated with; multiple regression will be a good option to predict the same with the help of the input variables i.e., income, neighbors, weight, length, count, loop (which are continuous in nature) and the output variable i.e., the label (either whitelisted i.e., not related with the ransomware addresses or the ransomware families that they are associated with) (Table 10).

Evaluation with Training Data: The correctness in predictability here is 80.59%, (as per the confusion matrix) which denotes that the model is quite healthy in classifying the addresses according to the ransomware groups (Table 11).

Evaluation with Test Data: The correctness in predictability here is 79.94%, (as per the confusion matrix) which denotes that the model is quite healthy in classifying the addresses according to the ransomware groups (Table 12).

Table 10. (a), (b), (c), (d), (e), (f). Multinomial Logistic Regression Output

	(Intercept)	length	weight	count	looped	neighbors	income
montrealCryptXXX	-0.109918436	0.002953426	-0.060880137	8.480941e-05	-0.0009524113	-0.090883739	-6.904804e-09
montrealDMALockerv3	-0.068851111	-0.015636180	-0.034131110	6.920852e-04	-0.0143662958	-0.116916321	-1.748937e-08
montrealSamSam	-0.031412066	0.014821938	-0.016053923	-8.944806e-04	-0.0020424270	-0.046610406	-4.179582e-08
montrealwannaCry	-0.007705343	-0.129979130	-0.001911949	3.005602e-03	-0.0003231190	-0.010949208	-4.017602e-08
paduaCryptowall	-0.030131433	0.018882654	0.119731140	-6.460968e-04	0.0006241057	-0.243884946	-1.672084e-09
princetonCerber	0.052942708	0.017699473	-0.048589084	-1.633947e-04	0.0001308873	0.299901706	-1.183969e-08
princetonLocky	-0.095260864	0.013602778	-0.068142797	-1.681623e-04	0.0004490959	-0.392618536	-1.232985e-09
white	0.548142449	0.049309560	-0.036875861	-1.182032e-02	0.0080942825	-0.009789657	5.747754e-10

(a)

	(Intercept)	length	weight	count	looped	neighbors	income
montrealCryptXXX	8.641574e-18	2.146056e-16	4.590439e-18	1.084107e-14	1.396354e-17	1.320149e-17	7.715618e-10
montrealDMALockerv3	2.215766e-17	8.277163e-16	1.071869e-17	4.153425e-14	7.168261e-18	3.537657e-17	2.087035e-09
montrealSamSam	8.889752e-17	1.283743e-15	3.924232e-17	6.295589e-15	4.087584e-17	1.401511e-16	5.486931e-09
montrealwannaCry	9.035377e-17	2.515027e-15	4.819882e-17	1.326722e-13	2.709934e-14	1.471664e-16	5.837168e-09
paduaCryptowall	1.476324e-17	2.949209e-16	7.461626e-18	5.351845e-15	1.597400e-15	2.200345e-17	2.026154e-10
princetonCerber	1.326542e-17	4.837659e-16	6.661034e-18	3.627013e-14	1.101737e-14	2.174917e-17	6.958674e-10
princetonLocky	1.279649e-17	6.172350e-16	5.584445e-18	2.844406e-14	1.189995e-14	2.107118e-17	1.962199e-10
white	5.152630e-17	5.916101e-16	2.312306e-17	2.414048e-16	6.010787e-17	8.366192e-17	3.331184e-11

(b)

```
> z
```

	(Intercept)	length	weight	count	looped	neighbors	income
montrealCryptXXX	-1.271972e+16	1.376211e+13	-1.326238e+16	7.822971e+09	-6.820702e+13	-6.884354e+15	-8.949126
montrealDMALockerv3	-3.107328e+15	-1.889075e+13	-3.184261e+15	1.666300e+10	-1.976253e+15	-3.304909e+15	-8.380006
montrealSamSam	-3.533514e+14	1.154588e+13	-4.090972e+14	-1.420805e+11	-4.996661e+13	-3.325726e+14	-7.617340
montrealwannaCry	-8.527972e+13	-5.168101e+13	-3.966795e+13	2.265434e+10	-1.192350e+10	-7.440017e+13	-6.882794
paduaCryptowall	-2.040977e+16	6.402617e+13	1.604625e+16	-1.207241e+11	3.907009e+11	-1.108394e+16	-8.252500
princetonCerber	3.991032e+15	3.658686e+13	-7.294523e+15	-4.504937e+09	1.188008e+10	1.378911e+16	-17.014284
princetonLocky	-7.444294e+15	2.203825e+13	-1.220225e+16	-5.912037e+09	3.773930e+10	-1.863296e+16	-6.283690
white	1.063811e+16	8.334808e+13	-1.594766e+15	-4.896473e+13	1.346626e+14	-1.170145e+14	17.254386

(c)

```
> p
```

	(Intercept)	length	weight	count	looped	neighbors	income
montrealCryptXXX	0	0	0	0	0	0	0.000000e+00
montrealDMALockerv3	0	0	0	0	0	0	0.000000e+00
montrealSamSam	0	0	0	0	0	0	2.597922e-14
montrealwannaCry	0	0	0	0	0	0	5.869083e-12
paduaCryptowall	0	0	0	0	0	0	2.220446e-16
princetonCerber	0	0	0	0	0	0	0.000000e+00
princetonLocky	0	0	0	0	0	0	3.306295e-10
white	0	0	0	0	0	0	0.000000e+00

(d)

```
Hosmer and Lemeshow test (multinomial model)

data:  trainingData$label, fitted(multinom_model)
X-squared = 2720.2, df = 64, p-value < 2.2e-16
```

(e)

```
CoxSnell  Nagelkerke   McFadden

0.3755068  0.4594581  0.2769807
```

(f)

Table 11. Confusion Matrix, for the training data (Multinomial Logistic Regression)

	multinom_Predicted					
	montrealCryptoLocker	montrealCryptxxx	montrealDMALockerv3	montrealSamSam	montrealWannaCry	paduaCryptoWall
montrealCryptoLocker	16	0	0	0	0	6
montrealCryptxxx	0	0	0	0	0	0
montrealDMALockerv3	0	0	0	0	0	0
montrealSamSam	0	0	0	0	0	0
montrealWannaCry	0	0	1	0	4	0
paduaCryptoWall	3	0	0	0	0	29
princetonCerber	2	0	0	0	1	0
princetonLocky	1	0	1	0	0	7
white	0	0	0	0	0	0

	multinom_Predicted		
	princetonCerber	princetonLocky	white
montrealCryptoLocker	15	3	383
montrealCryptxxx	8	2	51
montrealDMALockerv3	0	0	5
montrealSamSam	0	0	1
montrealWannaCry	0	0	1
paduaCryptoWall	12	8	190
princetonCerber	50	2	168
princetonLocky	14	17	108
white	0	0	4008

Table 12. Confusion Matrix, for the test data (Multinomial Logistic Regression)

	multinom_Predicted_validation				
	montrealCryptoLocker	montrealCryptxxx	montrealDMALockerv3	montrealSamSam	montrealWannaCry
montrealcryptoLocker	14	0	0	0	0
montrealCryptoTorLocker2015	0	0	0	0	0
montrealCryptxxx	0	0	0	0	0
montrealDMALockerv3	0	0	0	0	0
montrealFlyper	0	0	0	0	0
montrealNoobCrypt	0	0	0	0	0
montrealwannaCry	0	0	0	0	2
paduaCryptowall	1	0	0	0	0
princetonCerber	1	0	0	0	1
princetonLocky	1	0	1	0	3
white	0	0	0	0	0

	multinom_Predicted_validation			
	paduaCryptowall	princetonCerber	princetonLocky	white
montrealCryptoLocker	10	9	2	422
montrealCryptoTorLocker2015	0	0	0	3
montrealCryptxxx	0	11	0	40
montrealDMALockerv3	0	0	0	1
montrealFlyper	0	0	0	1
montrealNoobCrypt	0	0	0	2
montrealwannaCry	0	2	1	4
paduaCryptowall	20	17	12	161
princetonCerber	1	28	2	183
princetonLocky	10	9	19	115
white	0	0	0	4008

6 Conclusion

In this paper, a new data analytics-based framework has been proposed to detect and predict ransomware transactions that use Bitcoin. Using a topological data analysis-based approach and novel blockchain graph related features, it has been empirically shown that the Bitcoin addresses which are associated with fraudulent activities or/and are ransomwares themselves, could be easily predicted using sophisticated machine learning algorithms.

As a future work, it could be planned to combine the proposed approach with threat intelligence information (e.g., reports about the emergence of new ransomware families) to increase the prediction accuracy.

7   Suggestions and Recommendations

- It could be suggested that, the decision tree, random forest & binary logistic regression models should be used in order to classify the Bitcoin addresses into whitelisted & blacklisted labels whereas, the multinomial regression model will be used to identify the Ransomware family names amongst all the Bitcoin addresses.
- The Random Forest algorithm would be recommended over the Decision Tree algorithm because it has higher accuracy in terms of classifying the Bitcoin addresses.
- It is also being recommended to take as large as possible test dataset; so that the training dataset (sample) on which the model is been built, truly represents the entire dataset and makes the performance of the model much more realistic and acceptable.

8   Limitations

Few limitations that have been come across the way are:

- Due to the system capacity limitations, the test had to be done with a limited number of records only (10,234); if the training sample was increased, eventually the model accuracy would also have been increased.
- The main focus of this study was to build classification models in order to correctly classify the potentially fraudulent Bitcoin addresses or the ransomware addresses; hence still there is scope to optimize the model performances.
- The multinomial regression model trained in the current study, can only predict the labels or ransomware families that were been taken during building up the model (i.e., from the training dataset). It will not be able to predict any new ransomware family that has been emerged in the test dataset. Hence there is a scope to eliminate this limitation in future studies.

References

Akcora, C.G., Li, Y., Gel, Y.R., Kantarcioglu, M.: BitcoinHeist: topological data analysis for ransomware detection on the bitcoin blockchain. arXiv preprint arXiv:1906.07852 (2019)

Andronio, N., Zanero, S., Maggi, F.: HELDROID: dissecting and detecting mobile ransomware. In: Bos, H., Monrose, F., Blanc, G. (eds.) RAID 2015. LNCS, vol. 9404, pp. 382–404. Springer, Cham (2015). https://doi.org/10.1007/978-3-319-26362-5_18

Androulaki, E., Karame, G.O., Roeschlin, M., Scherer, T., Capkun, S.: Evaluating user privacy in bitcoin. In: Sadeghi, A.R. (ed.) FC 2013. LNCS, vol. 7859, pp. 34–51. Springer, Heidelberg (2013). https://doi.org/10.1007/978-3-642-39884-1_4

Conti, M., Gangwal, A., Ruj, S.: On the economic significance of ransomware campaigns: a Bitcoin transactions perspective. Comput. Secur. **79**, 162–189 (2018)

Di Battista, G., Di Donato, V., Patrignani, M., Pizzonia, M., Roselli, V., Tamassia, R.: Bit-coneview: visualization of flows in the bitcoin transaction graph. In: 2015 IEEE Symposium on Visualization for Cyber Security (VizSec), pp. 1–8. IEEE, October 2015

Goldsmith, D., Grauer, K., Shmalo, Y.: Analyzing hack subnetworks in the bitcoin transaction graph. Appl. Netw. Sci. **5**(1), 1–20 (2020)

Huang, D.Y., et al.: Tracking ransomware end-to-end. In: 2018 IEEE Symposium on Security and Privacy (SP), pp. 618–631. IEEE, May 2018

Liao, K., Zhao, Z., Doupé, A., Ahn, G.J.: Behind closed doors: measurement and analysis of CryptoLocker ransoms in Bitcoin. In: 2016 APWG Symposium on Electronic Crime Research (eCrime), pp. 1–13. IEEE, June 2016

Martín, A., Hernandez-Castro, J., Camacho, D.: An in-depth study of the jisut family of android ransomware. IEEE Access **6**, 57205–57218 (2018)

Maxwell, G.: CoinJoin: bitcoin privacy for the real world. In: Post on Bitcoin Forum, vol. 3, p. 110, August 2013

Meiklejohn, S., et al.: A fistful of bitcoins: characterizing payments among men with no names. In: Proceedings of the 2013 Conference on Internet Measurement Conference, pp. 127–140, October 2013

Nakamoto, S.: Bitcoin: a peer-to-peer electronic cash system. Decent. Bus. Rev., 21260 (2008)

Narayanan, A., Möser, M.: Obfuscation in bitcoin: techniques and politics. arXiv preprint arXiv: 1706.05432 (2017)

Ober, M., Katzenbeisser, S., Hamacher, K.: Structure and anonymity of the bitcoin transaction graph. Future Internet **5**(2), 237–250 (2013)

Paquet-Clouston, M., Haslhofer, B., Dupont, B.: Ransomware payments in the bitcoin ecosystem. J. Cybersecur. **5**(1), tyz003 (2019)

Rivera-Castro, R., Pilyugina, P., Burnaev, E.: Topological data analysis for portfolio management of cryptocurrencies. In: 2019 International Conference on Data Mining Workshops (ICDMW), pp. 238–243. IEEE, November 2019

Ruffing, T., Moreno-Sanchez, P., Kate, A.: CoinShuffle: practical decentralized coin mixing for bitcoin. In: Kutyłowski, M., Vaidya, J. (eds.) ESORICS 2014. LNCS, Part II, vol. 8713, pp. 345–364. Springer, Cham (2014). https://doi.org/10.1007/978-3-319-11212-1_20

Tschorsch, F., Scheuermann, B.: Bitcoin and beyond: a technical survey on decentralized digital currencies. IEEE Commun. Surv. Tutor. **18**(3), 2084–2123 (2016)

A Taxonomy and Survey of Software Bill of Materials (SBOM) Generation Approaches

Vandana Verma Sehgal⬤ and P. S. Ambili$^{(✉)}$ ⬤

REVA University, Bangalore 560064, India
ambili.ps@reva.edu.in

Abstract. The software supply chain has been there for a very long time, and so do the issues, risks and vulnerabilities associated with it. In the past few years, Supply chain attacks are even more, as the authors have started using more open-source software, code, dependencies, and libraries in the code. The open-source code is there to help us out, however, it does bring third-party risks with it. If not fixed with it. Software Bill of Materials (SBOM) with the right set of configurations and automation is one of the possible solutions. Staying up to date with the latest security and SBOM updates is a tedious task. Every day, there are new exploits created and new patches released. The intent of the study is to share the insights and results of the SBOM implementation review. The authors reviewed different Software Bill of Materials like OWASP Cyclone Dx and SPDX by Linux Foundation. It is imperative to learn about algorithms and implementations.

Keywords: Software Bill of Materials · Cyber Security · Automation · Supply Chain · Software Supply Chain Security

1 Introduction

The authors as the tech world cannot function without open-source software (OSS), but despite extensive research on specific (typically central) projects, there is little knowledge of the OSS ecosystems [14]. Every day the authors are hearing about new attacks creeping up with the software supply chain. As much as Software Supply chain helps in getting the applications and software developed faster. There are risks associated with the insecure supply chain or unwanted addition of software piece in the whole chain. Software Bill of Materials is useful for protecting automation and the supply chain ecosystem. The process of inventory management includes SBOM. The study can protect it if the authors have the inventory.

Businesses frequently experience hacking. Creating an SBOM is one way to prevent security breaches from happening. Automation is a crucial component of any SBOM; picture it as involving robots in the management of your company. Automation may be used for a variety of purposes in your business, depending on your needs and objectives.

A thorough study was done on the following in this work:

- The benefits of using SBOMs for security purposes, such as identifying vulnerabilities and tracking components

© The Author(s), under exclusive license to Springer Nature Switzerland AG 2024
S. Dhar et al. (Eds.): AGC 2023, CCIS 2008, pp. 40–51, 2024.
https://doi.org/10.1007/978-3-031-50815-8_3

- The challenges and limitations of SBOMs in terms of security
- Best practices for creating and maintaining SBOMs for security purposes.
- Checked the Case studies or examples of organizations that have successfully used SBOMs to improve software security.
- Comparison of different approaches to SBOM creation and management
- Future directions for research and practice in SBOM security
- How to secure the SBOMs

The work explores SBOMs and other security practices, such as supply chain risk management, threat modelling, and vulnerability management [3]. The aim of the study is to aggregate various implementations of Software Bill of Materials across various industrial implementations and to identify the variations. To that end, the authors carried out a Systematic Literature Review that the academic community and industry that is interested in this subject can use.

There are six sections included in the paper are: Sect. 1 is the Introduction of the paper. Software Bill of Materials is covered in the following Sect. 2. The literature review process is highlighted in Sect. 3, the conclusions the authors came to from it are discussed in Sect. 4, and the final considerations are discussed in Sect. 5.

2 Software Bill of Materials

A "Software Bill of Materials" (SBOM) is an inventory containing all the software's used in a product or that make up a product starting with the name of the package, version of the package and vendor information. Whenever there is something that should be followed, there must be standard for the same. The authors checked a lot of places and found that there are two or three standards which are very influential in the industry out of which Cyclone Dx, SWID and SPDX are the most prominent once [13].

The Software Package Data Exchange, or SPDX, is an open standard for exchanging information about the components, licenses, copyrights, and security references that make up a software bill of materials. By offering a standard format for businesses and communities to share crucial data, SPDX reduces redundant work, streamlining and improving compliance. CycloneDx offers cutting-edge supply chain capabilities for lowering cyber risk. List the services and components of the software, along with their dependencies. The SWID standard specifies a lifecycle in which a SWID tag is added to an endpoint during the installation of software and removed during the uninstallation of software [21].

While all the formats try to convey a similar set of information, each of them has the data represented in different chucks and the focus of the bill of material is slightly different for each of the format. For ex. SPDX is more about information sharing and focuses on software license compliance, CycloneDx aims to create aims to create the most extensive specification for creating Software bill of materials. It can be automated easily. SWID tags are designed to make it simpler to find, recognize, and contextualize software throughout the software lifecycle, assisting enterprises in building accurate software inventories. The standard is supported by NIST and is a subset of the larger ISO IT asset management standard.

Cyclone DX and SPDX both defines the data structure very differently. SPDX could be a considered a good choice for human readable format whereas excellence CycloneDx format specification makes it's an excellent automation target. The authors can write code which makes it easy to use for machines. SWID can be used in the pre-installation phase that can be used in the form of TAR, ZIP file, executable file.

SPDX can be used in different formats like RDFa, .xlsx, .spdx and expand into other formats such as .xml, .json, and .yaml. On the contrary, CycloneDX can be used or represented in formats like XML, JSON. SPDX is an recognized as an International Organization for Standardization (ISO) whereas CycloneDx is a community initiative. Consumers are the organizations and tools which take input from the SBOM tools and provide the relevant output in the form of vulnerabilities. The authors need to understand how these SBOM's can be created.

Creation of Software Bill of Materials: -

SPDX: The BinTray SPDX Tools Java repo under the appropriate release has the SPDX Tool binaries available for download. The package can be found in Maven Central as well (organization org.spdx, artefact spdx-tools).

CycloneDx: CyclonDx can simply be called and downloaded via npm package install script npm install -g @cyclonedx/bom. The tool will automatically produce the bill of materials in XML format. Using cyclonedx-bom -o sbom.json, the authors can override that and instruct it to create a JSON-format SBOM instead.

XML format CycloneDx:

```
{
  "bomFormat": "SBOM",
  "specVersion": "2.1",
  "serialNumber": "urn:uuid:3e67188-792b-bc0a-a67575785bnbb79",
  "version": 1,
  "components": [
    {
      "type": "library", "name": "acme-library", "version": "2.0.0"   } ] }
```

SWID Creation: SWID and CoSWID tag creation can be performed via Java API. The API enables the creation of tags in the XML-based SWID and CBOR-based Constrained SWID (CoSWID) formats in Java programmes. The builder patterns offered by this library can be combined to create a SWID or CoSWID tag (Fig. 1).

After reviewing the different SBOM standards deeply, the authors realized, to optimize the SBOM process. The survey need to dig deeper and need to understand the third-party ecosystem. For this purpose, the authors then analyzed many dependencies and third-party binaries which led us to a few questions to ponder upon: -

Q1: Are all the vulnerabilities that are identified captured in some systems?
Q2: Can the authors predict a supply chain risk without analyzing the source code of the project?
Q3: What are the challenges in adopting automated package upgrades?

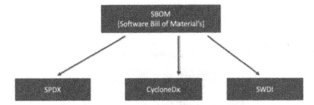

Fig. 1. Software Bill of materials Creation Process

3   Literature Review

There has been some significant work which has been done in the area of software supply chain issues. The authors tried to understand and figure out the research work which has been done so far in the space.

To start the research with different phases of review:

- Initializing the Review
- Performing the Review
- Drawing the conclusions

These phases helped in narrowing down the research and understanding what is relevant information/data and what the authors need to discard or not use further.

Initializing the Review: In this phase, the authors started the research with certain set of things the authors follow a certain path for the research.

- Formulating the research questions
- Kind of activities to perform.
- Data/information to be kept or removed.
- Extracting the relevant result from the data

Understanding that a problem exists and the authors need to solve it is always a first and important step. This helps as a stepping stone. After identifying the problem around open source dependencies and Software Bill of Materials around issues and risk.

With SBOM keyword search, The initial survey included 30 odd papers out of which the authors selected five papers that highlighted the Software Bill of Materials, the importance of using SBOM's, and securing the organisation by tracking third-party dependencies. Cybersecurity Automation was one of the keywords which came up with thousands of papers which the authors narrowed down to 20 papers relevant to open-source dependency issues to Supply chain risks. To perform the review, the authors followed a phased approach in using the relevant keywords and adding the different keywords to get the related research.

The ratio of paper selection was a very interesting mix of containing the overview of supply chain issues in the ecosystems and promptly highlighting (Fig. 2):-

- Vulnerability timeline
- Framework
- Automation

- Vulnerability Identification and tracking

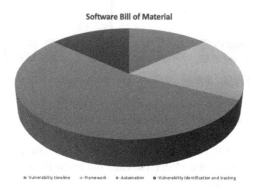

Fig. 2. Literature Selection Proportion

The authors selected the papers from different sources like Journals, websites, research reports, conference papers etc. The review process was conducted based on the following open query concerns (Fig. 3):

Q1: How the organization should use the Software bill of materials to keep an eye on dependency issues?
Q2: How can organizations understand the posture of their open-source components and track the risk associated?
Q3: What businesses can do to evaluate and reduce the risks associated with the usage of third-party vulnerable components?

Fig. 3. SBOM Literature Selection Process

In addition to enabling permanent recording of the action and timing of data provision requests, storing metadata that may identify smart contracts on the blockchain also opens up the possibility of encoding quality of service requirements and enabling automatic payment for data provision requests.

Best Practices for Creating and Maintaining SBOMs for Security Purposes: Make SBOM creation and maintenance a priority: Ensuring that SBOMs are complete and accurate should be a high priority for any organization that is serious about software security. This may require dedicated resources for SBOM creation and maintenance, making it a key part of the software development and procurement process. Use automated tools to create and maintain SBOMs: There are a variety of tools available that can help automate the process of creating and maintaining SBOMs. These tools can help ensure that SBOMs are complete and up-to-date and can save time and effort compared to manual processes. Use a consistent, standardized format for SBOMs: Using a consistent, standardized format for SBOMs can help ensure that they are easy to read and understand and can facilitate integration with other tools and processes. Regularly review and update SBOMs: SBOMs should be regularly reviewed and updated to ensure that they are accurate and up-to-date. This may involve tracking changes to the software and its components, as well as identifying and addressing any vulnerabilities or other security issues. Securely store and protect SBOMs: SBOMs should be treated as sensitive, confidential information and should be securely stored and protected to prevent unauthorized access or tampering. Use SBOMs as part of a broader security strategy: SBOMs should be viewed as just one part of a broader security strategy and should be integrated with other security practices, such as supply chain risk management and vulnerability management.

Comparison of Different Approaches to SBOM Creation and Management: There are a variety of different approaches to creating and managing software bill of materials (SBOMs). Some common approaches include:

Manual creation and maintenance: In this approach, SBOMs are created and maintained manually, often through the use of spreadsheet or database software. This can be time-consuming and prone to errors, but may be suitable for organizations with small software portfolios.

Automated Tools: There are a variety of automated tools available that can help create and maintain SBOMs. These tools can scan software and automatically identify components and their associated metadata, and can help track changes and updates to the software over time.

Open-Source Initiatives: A number of open source initiatives, such as the Open Chain Project, have been established to promote the use of SBOMs and provide tools and resources for creating and managing them. These initiatives can help organizations adopt best practices and standardize their SBOM processes.

Commercial Solutions: There are also a number of commercial solutions available that can help create and manage SBOMs. These solutions may offer a more comprehensive set of features and may be suitable for larger organizations with complex software portfolios. When comparing different approaches to SBOM creation and management, organizations should consider factors such as the size and complexity of their software portfolio, the level of automation and support needed, and the resources and budget available (Table 1).

Table 1. Key findings from Literature review

Author and Reference Link	Algorithms/Methodology	Merits	Limitations
Benjamin Rombaut et al. https://dl.acm.org/doi/10.1145/3522587	Discussion on Dependency bot and their methodology to identify vulnerabilities	Only bugs can be raised, the author still needs to perform the update	Limitations in the ecosystem if there is no ci/cd pipeline setup. Needs to manually intervene and update the third-party dependency
Nasif Imtiaz et al. https://doi.org/10.1145/3475716.3475769	Comparative study of different third-party dependency tools. Developers can deploy more than one tool to detect third party dependencies	Tools are limited by programming languages and vulnerability database	There is no common or publicly available vulnerability database from where the vulnerability details can be derived
Kui Liu et al. https://ieeexplore.ieee.org/document/9724801	Supply chain issues multiply as we look down on the dependency chain	People focus on primary language files and can miss the secondary language files	If you have a dependency tool track onmly java then the other languages bug can't be detected. All the tools are not universal, different tools focus on different languages
Xinyuan Wang https://ieeexplore.ieee.org/document/9652901	Detection of specific supply chain attacks is possible using packet and traffic analysis/network flow analysis	The issue pertaining with supply chain attacks and the technical difficulties in identifying them in software Empirical validations is pending for the proposed detection	There has been theoretical attempt to identify and protect against supply chain attacks using network flow analysis
Nasif Imtiaz et al. https://ieeexplore.ieee.org/document/9792380	Empirical study on fixed to release delay and associated impact on reliant projects	Security releases may also bundle breaking changes	Do not segregate security releases as separate releases. The security releases should be atomic and non-breaking
Duc-Ly Vu et al. https://ieeexplore.ieee.org/document/9678526	Identification of source code corresponding to the python package and its health check	Reliance on package repository format for source indication	Vulnerability identification is a very critical concern with the closed source packages
William Enck et al. https://ieeexplore.ieee.org/document/9740718	Highlighting the top five issues in the supply chain ecosystem	Understanding the risk can help in securing the supply chain pipeline to an extent and avoiding security risks	Good overview of supply chain issues in the ecosystems
Filipe R. Cogo et al. https://ieeexplore.ieee.org/document/9786016/	The paper talks about how we can customise dependabots. It is a study on where do the developer needs the customisation and most common dependency bots customisation are done for noise reduction	Over zealous noise reduction can result in security lapse	While automation definitely helps, over reliant on automation can result in too much noise. If noise reduction methods are applied, it may result in security lapse

(*continued*)

Table 1. (*continued*)

Author and Reference Link	Algorithms/Methodology	Merits	Limitations
Thomas Zimmermann2 et al. https://ieeexplore.ieee.org/document/9794068	By empirically examining npm package metadata, one can measure supply chain weak link signals to stop future supply chain attacks	framework for npm supply chain weak link signal prioritisation and quantification	Suggesting a framework for quantifying the risk for meta data parameters. People are coming with newer frameworks via different approach
Gabriel Ferreira et al. https://doi.org/10.1109/ICSE43902.2021.00121	Proposal of a lightweight permission system, enforcing least privilege design in npm	It can not protect modules that legitimately needs higher permission and has run time overheads	To reduce the risk on system in case of malicious or vulnerable package
Jing Zou et al. https://dl.acm.org/doi/10.1145/3349341.3349395	Self-test solution with security in mind could be useful for overall health of the ecosystem	The organisations needs to enforce this	New framework
Alexandre Decan et al. https://doi.org/10.1145/3196398.3196401	An empirical analysis of almost 400 security reports from the over 610k JavaScript packages in the npm dependency network over a 6-year period Vulnerabilities	A third of the vulnerabilities are fixed at (or before) the date of discovery, while half require more time but are fixed before the date of publication	Study on vulnerability fixing timeline

4 Outcome of the Review Process

William et al. [10] provided a nice overview of the threat landscape for the third-party libraries. The paper talks about Policy challenges like getting the industry to participate in adopting the Software bill of materials to leveraging the SBOM for security. It also throws light on challenges of updating vulnerable dependencies. There is a need to secure the build process and create a trustable & verifiable software bill of material.

Nasif et al. [5] identified that developers avoid security critical nature of the patches. They often release security bug fixes which are non-atomic and have breaking changes. This generally results in hesitation to upgrade the third party dependencies at a fast pace. Alexadre et al. [1] confirmed this assumption as they identified a long lead time between vulnerability bug fix or dependency update in the reliant project. A third of the vulnerabilities are fixed at (or before) the date of discovery, while half require more time but are fixed before the date of publication. The remaining 15% or so are regarded as high risk because they are either not fixed at all or fixed only after the vulnerability is made public.

While people are drawing interesting conclusions. There is another set of research, trying to create set of frameworks around solving this problem as mentioned below:-

Jing et al. [16] try to address the problems of security bugs in the open-source ecosystems by creating a framework to detect and fix the bug at the initial stage by incorporating security practices in the development as run time phase itself. However, due to the requirement to interact with the source code to the software, this framework can only be applied by the organization or the individual owner of the code.

Packet and Network analysis could pave the way for detecting known supply chain attacks. This hypothesis is experimented by Xinyuan Wang [4] and they came up with a framework for the same. Empirical validations is pending for the proposed detection.

Gabrielle at al. [6] talks about the usage and reusage of software packages in the modern ecosystems like npm, python, go etc. While the authors use such packages, it is utmost important to keep the packages up to date to avoid using vulnerable components. They attempt to prevent supply chain attacks by implementing a least privilege framework using the lightweight permission system. However, the framework will only work if all parties agree to implement it and the framework has a known limitation where it can prevent abuse via packages which legitimately need higher permissions.

In an attempt to identify simple indicators to a possible supply chain risk, Nusrat et al. [15] worked on identifying weak link signals using the metadata of the package such as email address domain, busyness of maintainer, number of contributors to the open source packages, existence of installation script and age of the last commit.

Benjamin et al. [17] talks about the automated detection of these third party dependency issues via dependency bots. They further explores the limitations around automated pull requests and vulnerability fix. However, there are limitations with the dependency bots if there is no proper CI/CD pipeline setup. The authors will have to manually intervene and test the issues which further leads to spending more time. On the contrary, Filipe et al. [8] identified the other end of the spectrum where due to too much automation, developers end up tweaking the dependency bots to reduce noise and if this attempt goes overboard, it results in security lapse. Hence, it is essential that any attempt to automate should consider the possible impact on the maintainer.

On one hand, third party risks are a problem where in SBOM's are a first good step. Nasif et al. [5] talks about SBOM's are great however they need to be fed into a vulnerability database to find the bugs, The key ingredient to work for this is vulnerability database. Also, it has been identified that there is no single place or platform which captures all the vulnerability databases. This means the vulnerability logic varies from each software depending on the accumulated database.

As per Kui et al. [18] Also, Supply chain issues and risks multiply as the authors look down on the dependency chain. People focus on primary language files and can miss the secondary language files. If you have a dependence tool that can track only Java then the other languages bug like npm, python etc. which are reliant on the dependency can go undetected. All the tools are not universal, different tools focus on different languages.

On one side Yuxing [19] et al. has created a world of code environments which analyze the interdependencies of the open-source ecosystems. They mentioned World of Code to be used as a steppingstone for a flexible infrastructure that can be updated and expanded to support tools and research that rely on version control information from all open source projects is discussed along with some of the research issues that call for such a broad scope. The system could be halted if the application ends up using the closes source packages which was a major concern for Duc-Ly Vu et al. [8] while they attempted to perform at-scale vulnerability identification using the package source metadata from the Python package repository.

Vulnerability identification is a very critical concern with closed-source packages. The authors always feel that the code written by us is inherently secure. However, as much as there are bugs in open-source projects, the same can be seen in closed-source if not written properly.

Presented below are the evaluation the initial research questions that the authors had and the ascertained answer based on the literature review.

Q1: Are all the vulnerabilities that are identified captured in some systems?

As identified by [5], there is no uniform publicly accessible database. This poses a risk for the SBOM consumption as every consumer must rely on their own vulnerability databases.

Q2: Can the authors predict a supply chain risk without analyzing the source code of the project?

While having access to source code will speed up the vulnerability identification process as identified by [8], Closed source packages are at a disadvantage, however, attempts by [15, 19, 20], do confirm that some conclusions can be derived by leveraging the project metadata.

Q3: What are the challenges in adopting automated package upgrades?

Certain level of automation can be achieved by leveraging dependency bots. However, the usage of dependency bots should be considerate to the developers as in [8], Developers might end up over optimize the code (to reduce noise) resulting in security lapse while giving a false sense of security. Furthermore, As identified by [5] packages need to have non-breaking security patches. The automation needs to be mindful about the nth level dependencies as they may not be in the same language's framework [18].

With the selection process, the authors realized that it is important to understand the accountability and traceability [21, 23] of the content and assets. It also varies from domain to domain. It is important to also narrow down on the Models that the authors wish to apply while building the SBOM, what languages like R, Python, Go or GraphQL etc. will be used for the schema. Are the authors using machine learning algorithms to test out the data. There are some wonderful models used as part of SBOM's.

During the literature review process, The authors narrowed down the SBOM selection process, the creation of a BoM for each system offers a mechanism to archive the ecosystem for each experiment by looking beyond the data and taking into account additional contributing factors such as the software and hardware that produces or manages the data, licenses that govern the use and sharing of the data, and policies that contributed to the generation of the data.

The authors find the key applications of SBOM's could be with healthcare industry, Technology Industry, Airline Industry and Academics [22]. SBOM selection process is critical and using the right technology to build that, the authors will be helping the academic institutions to build the software for the Institutions and helping the students. Every time the system operates, the BoM is converted into Bill of Software, adding a dynamic and traceable perspective to the static parts list. As a result, the data inputs, data outputs, and any artefacts used or created by the system may all be archived, easily recognized, and traced back to their original sources.

5 Conclusion

The study wishes to contribute research to enhance the processes around the supply chain to detect the software's and third parties used in it. That will directly and indirectly help eradicate the attacks on organizations to a great extent. The study attempts to create solutions which can help self-updating cloud solution that automatically identifies and fixes any vulnerabilities in the code, can help you stay more secure and out of danger! Organizations might leverage SBOMs to proactively identify and address vulnerabilities and potential security risks within their software supply chains and in future SBOM could lead to better risk management and improved overall security postures.

References

1. Decan, A., Mens, T., Constantinou, E.: On the impact of security vulnerabilities in the npm package dependency network. In: Proceedings of the 15th International Conference on Mining Software Repositories (MSR 2018), pp. 181–191. Association for Computing Machinery, New York (2018). https://doi.org/10.1145/3196398.3196401
2. Zhang, G., Xu, Y., Hou, Y., Cui, L., Wang, Q.: Cyber-security risk management and control of electric power enterprise key information infrastructure. IEEE Mol. **8** (2022)
3. Yan, D., Liu, K.: 2021 IEEE 21st International Conference on Software Quality, Reliability and Security, p. 6 (QRS) (2021)
4. Wang, X.: IEEE, On the Feasibility of Detecting Software Supply Chain Attacks, p. 6 (2021)
5. Imtiaz, N., Khanom, A., Williams, L.: Open or Sneaky? Fast or Slow? Light or Heavy?: Investigating Security Releases of Open Source Packages, p. 12. IEEE (2022)
6. Ferreira, G., Jia, L., Sunshine, J., Kästner, C.: Containing malicious package updates in npm with a lightweight permission system, p. 16. ACM (2021)
7. Cogo, F.R., Hassan, A.E.: An empirical study of dependency downgrades in the npm ecosystem. IEEE Trans. Softw. Eng. **47**(11), 2457–2470 (202). https://doi.org/10.1109/TSE.2019.2952130
8. Vu, D.L., Pashchenko, I., Massacci, F., Plate, H., Sabetta, A.: Towards using source code repositories to identify software supply chain attacks. In: Proceedings of the 2020 ACM SIGSAC Conference on Computer and Communications Security (CCS 2020), pp. 2093–2095. Association for Computing Machinery, New York (2020). https://doi.org/10.1145/3372297.3420015
9. Pashchenko, I., Vu, D.-L., Massacci, F., Plate, H., Antonino: LastPyMile: identifying the discrepancy between sources and packages. In: IEEE European Symposium on Security and Privacy Workshops (EuroS&PW), vol. 7 (2021)
10. Enck, W., Williams, L.: Top Five Challenges in Software Supply Chain Security: Observations From 30 Industry and Government Organizations. North Carolina State University (2021)
11. Singi, K., Jagadeesh Chandra Bose, R.P., Podder, S., Burden, A.P.: Trusted software supply chain. In: Proceedings of the 34th IEEE/ACM International Conference on Automated Software Engineering (ASE 2019), pp. 1212–1213. IEEE Press (2020). https://doi.org/10.1109/ASE.2019.00141
12. CISA gov. https://ntia.gov/files/ntia/publications/ntia_sbom_energy_jan2021overview_0.pdf. Accessed Nov 2021
13. https://www.erp-information.com/spdx-vs-cyclonedx#SPDX_vs_CycloneDx. Accessed August 2021
14. Announcing the 2022 State of Open Source Security report from Snyk and the Linux Foundation. https://snyk.io/reports/open-source-security-2022/. Accessed March 2022

15. https://snyk.io/blog/announcing-2022-state-of-open-source-security-report/. Accessed January 2022
16. Zimmermann, T., Godefroid, P., Zahan, N., Murphy, B., Maddila, C., Williams, L.: What are Weak Links in the npm Supply Chain? https://ieeexplore.ieee.org/document/9794068
17. Zou, J., Zeng, W., Zhao, Y., Liang, R., Cai, L., Zhao, Y.: Research on secure stereoscopic self-checking scheme for open-source software. In: Proceedings of the 2019 International Conference on Artificial Intelligence and Computer Science (AICS 2019), pp. 158–162. Association for Computing Machinery, New York (2019). https://doi.org/10.1145/3349341.3349395
18. Rombaut, B., Cogo, F.R., Adams, B., Hassan, A.E.: There's no such thing as a free lunch: lessons learned from exploring the overhead introduced by the greenkeeper dependency bot in npm. ACM Trans. Softw. Eng. Methodol. **32**(1), Article no. 11, 40 p. (2023). https://doi.org/10.1145/3522587
19. Liu, K., Liu, Z., Yan, D., Niu, Y., Liu, Z., Bissyande, T.F.: Estimating the attack surface from residual vulnerabilities in open source software supply chain. IEEE Xplore (2021). https://ieeexplore.ieee.org/document/9724801
20. Ma, Y., Bogart, C., Amreen, S., Zaretzki, R., Mockus, A.: World of code: an infrastructure for mining the universe of open source VCS data. In: 2019 IEEE/ACM 16th International Conference on Mining Software Repositories (MSR), Montreal, QC, Canada, pp. 143–154 (2019). https://doi.org/10.1109/MSR.2019.00031
21. Abraham, B., Ambili, P.S.: An enhanced career prospect prediction system for non-computer stream students in software companies. In: Chatterjee, P., Pamucar, D., Yazdani, M., Panchal, D. (eds.) Computational Intelligence for Engineering and Management Applications. LNEE, vol. 984, pp. 811–819. Springer, Singapore (2023). https://doi.org/10.1007/978-981-19-8493-8_60
22. Guidelines for the Creation of Interoperable Software Identification (SWID) Tags. https://www.govinfo.gov/app/details/GOVPUB-C13-626f2b896b2c32911a45026542e29268. Accessed Jan 2022
23. Ambili, P.S., Abraham, B.: A predictive model for student employability using deep learning techniques. ECS Trans. **107**(1), 10149 (2022). https://doi.org/10.1149/10701.10149ecst
24. Barclay, I., Preece, A., Taylor, I., Verma, D.: Towards traceability in data ecosystems using a bill of materials model (2019). https://doi.org/10.48550/arXiv.1904.04253

Find Your Donor (FYD): An Algorithmic Approach Towards Empowering Lives and Innovating Healthcare

Tamoleen Ray(✉) [ID]

University Institute of Technology, University of Burdwan, Bardhaman, West Bengal, India
tamoleen@gmail.com

Abstract. In the past few decades, we have seen an exponential increase in demand in health care services. But yet we are limited to deliver these services to the ones in need. Find Your Donor (FYD) will bridge the gap between interested blood and organ donors and needy patients for blood, and other emergency services through User Interfaces, deployed through secure ML pipelines where the data is stored securely in a cloud-based system. Cloud computing is a new way of delivering computing resources and services. Experts around the world believe that it can improve health care services, benefit health care research, and change the face of health information technology. Health-care industries need to be integrated with ML Algorithms for efficient use. Let us suppose a person meets with a serious road accident. In this case of an emergency, rare blood groups like say O negative donors are not always available. Here we implement machine learning techniques to reach out for interested acceptors to a particular donor. Again, this will also help patients suffering from diseases like Thalassaemia who need blood of a particular group at regular intervals.

Keywords: Machine Learning · Deep Neural Networks · Collaborative filtering approach · Digital India · Biometric Authentication · Blood Donation · Blood Bank-Management System

1 Introduction

According to the latest published report by World Health Organization (WHO) on Blood safety and availability, Of the 118.5 million blood donations collected globally, 40% of these are collected in high-income countries, home to 16% of the world's population. In low-income countries, up to 54% of blood transfusions are given to children under 5 years of age; whereas in high-income countries, the most frequently transfused patient group is over 60 years of age, accounting for up to 76% of all transfusions. Based on samples of 1000 people, the blood donation rate is 31.5 donations in high-income countries, 16.4 donations in upper-middle-income countries, 6.6 donations in lower-middle-income countries and 5.0 donations in low-income countries. An increase of 10.7 million blood donations from voluntary unpaid donors has been reported from 2008 to 2018. In total, 79 countries collect over 90% of their blood supply from voluntary unpaid

S. Dhar et al. (Eds.): AGC 2023, CCIS 2008, pp. 52–61, 2024.
https://doi.org/10.1007/978-3-031-50815-8_4

blood donors; however, 54 countries collect more than 50% of their blood supply from family/replacement or paid donors.

The need of rare blood groups for persons in need increases as days pass by. At times even the blood banks cannot suffice the need of the commons. Here comes the role of Find Your Donor (FYD) application which helps us to bridge the gap between the interested donors and the patients in need. The salient features of this app are: Unique access; Each user has to register and login further with his/her own personal credentials. Contact Details: The location or address should be entered by the user and would be used for contact purposes. Apart from this it will enable direct contact of the patient with the donor. Round the clock service at your hands: It will establish 24 × 7 connection for anyone in need. It is seen at times that many patients in hospitals do not get blood of their blood group due to: i) very rare blood groups. ii) no donors at the hospital or local premises. Again, people suffering from diseases like Thalassaemia, Blood Cancer etc. suffer from these problems every month. Hence required measures need to be taken. Here FYD bridges the gap between patients in need and the interested blood donors.

2 Literature Survey/Previous Related Works

In past few years many organizations like the Robinhood Army, Smile Foundations have moved towards bridging similar gaps but this application is one of its own kind.

This application is one of its unique as it focuses on medical and healthcare benefits provided to the persons in need.

Previous related surveys include Robin Hood Army: Founded on 26 August 2014, the Robin Hood Army is a volunteer-based NPO that works to get surplus food from restaurants to the less fortunate sections of society in cities across India and 10 other countries. The organization consists of over 218,912 volunteers (approximately) in 401 cities, and has served food to over 118.46 million people so far. The organization reaches out to homeless families, orphanages, old-age homes, night shelters, homes for abandoned children, patients from public hospitals, etc. Their prime motto is Fighting Hunger by collecting surplus food from restaurants and the community to serve less fortunate people; Smile Foundation: An NGO which focuses on child education for poor children, healthcare for families, skills training and livelihood for youth, and community.

While designing Find Your Donor: FYD China's Blood Management System [6] has produced some greater insights and is one of the notable case studies. Some of the steps including maintenance of blood-antigen data across every notably small population density, Ensuring and enforcing special guidelines for decreasing the spread of blood related diseases.

There stands a limitation to literature survey in Find Your Donor since this use-case is of a very high novelty. Similar steps may have been taken by NGO's or Blood banks of various nations including above mentioned organizations but FYD is an algorithmic approach towards solving a perfect match for rare blood donors or organ donors in case of emergency is one of the unique in its kind [1,4,5,7,11].

3 Proposed Methodology

Firstly, we design a donor to acceptor mapping which will further help us generate the required python function for the desired task.

Keeping in mind the architecture and the algorithmic design in which we desire to perform the task we continue as per the proposed methodology.

3.1 Donor to Acceptor Matrix

The donor to acceptor matrix is (Fig. 1.) which is constructed for constructing a mapping from Donor to Acceptor i.e., a one-to-many function is constructed based on the (Fig. 2.)

CAN RECEIVE GROUP								
This will be a binary matrix of 8X1 Each entry 0 or 1								
	O+	O-	A+	A-	B+	B-	AB+	AB-
O+	1	1						
O-		1						
A+	1	1	1	1				
A-		1		1				
B+	1	1			1	1		
B-		1				1		
AB+	1	1	1	1	1	1	1	1
AB-		1		1		1		1

(Recipient's Blood Group)

Fig. 1. Donor to Acceptor Matrix

Type O negative (O−) is a universal donor whereas, type AB positive (AB+) is a universal recipient/acceptor. This suggests a major role in our proposed methodology. A sketch of the required function mapping is shown in Fig. 3. Where the domain set is comprised of the recipient's blood group and the co-domain set is comprised of the donor's blood group. It is clearly seen that O negative, B negative, etc. are rare blood groups and hence are more concerned for 'Find Your Donor' application.

3.2 Emergency Flag Mechanism

Based on Figs. 1, 2 and 3 we group the bloods into three categories which will act as the parent attribute for transmitting emergency signals via 'Find Your Donor' App. Flag 1: Flag 1 is raised in stage 1 when there is very less or mild emergency i.e., if the blood group of the patient falls into categories of AB positive (AB+), A positive (A+) or B positive (B+). Flag 2: In case stage 2 arrives a pool of contacts are driven which may include a search over a 100 km radius for a perfect match. If the blood group of the patient falls into categories of AB negative (AB−), A negative (A−) or B negative (B−). Flag 3: In case 3rd stage flag is raised the donors are contacted right away, volunteers or any associates of Find Your Donor are contacted and best possible match are found as soon as possible. Blood group of these categories are O negative (O−) and O positive (O+).

3.3 Architecture

The proposed mechanism of Find Your Donor consists of a cloud-based system where data is securely stored and retrieved upon two stage authentications. The search mechanism works on Dijkstra's Shortest Path Algorithm. Finally, each of the Donor's data

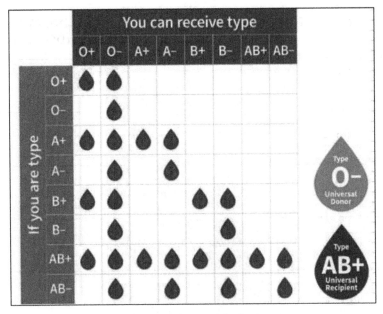

Fig. 2. Blood group Distribution

is driven into a recommendation engine through push mechanism using Collaborative Filtering technique.

Description

Find Your Donor (FYD) will be integrated through cloud servers. This is a client server network which will follow star topology. At first each user/client will be recognized with their unique id and password through which query can be placed in case of a receiver or patient's side and an existing query can be received from the donor side interface. In case of receiver side, the data is first retrieved and then filtered based on the tuned parameters defined during the ML workflow. The architecture will consist of an admin side also which will take care of the processes by various users globally connected in a WAN. After any changes made either from the user/admin side there will be an additional layer of authentication which if true then the changes will be saved in the cloud server. Thus, any manual write operation will be proceeded via a two-step authentication medium as shown in Fig. 4.

ML Algorithmic Workflow

Firstly, each blood group is mapped onto as per the Donor to acceptor matrix as discussed above and a function is created for the same in Python 3. Secondly, the model is trained using two metrics required blood groups (req_bl_grp) and distance from acceptor (dis_acc). Next the parameters bl_grp and dis_acc are tuned. Further, the model is split into train/dev/test sets and the ML model is further trained. Finally using Collaborative Filtering approach [3], we complete the desired algorithm.

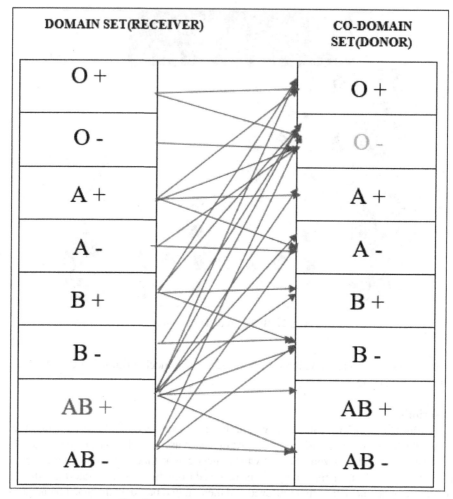

Fig. 3. Recipient to Donor Function Mapping

Classifying Healthy Subjects from Interested Donors. In FYD while registration medical related documents are a must and here the above-mentioned transfer learning architecture (Fig. 5.) comes into play by Classifying Healthy Subjects from Interested Donors. The above-mentioned transfer learning neural network [9] is a prototype for the actual network to be followed in FYD's algorithmic workflow. In a transfer learning mechanism, the same neural network can be used to perform similar tasks. The classification network will be designed for a single site or area which on applying transfer learning will be adaptable to perform similar classification on the constraints of a particular geographic location, particular time frame of a day, temperature, etc. Now the model is ready to perform globally where the geographical constraints may change but a patient's search for a healthy donor is optimized by transfer learning. At first the model will be trained using the training sets available. The input layers here represent the interested

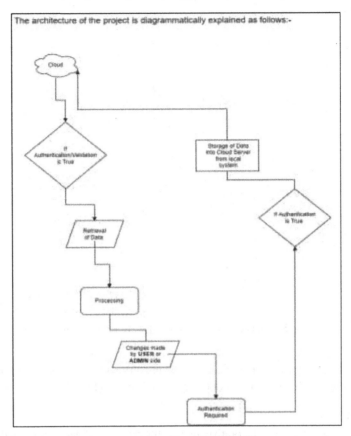

Fig. 4. Client-Server Cloud Based Architecture.

donors from various sites and the output layer consists of the healthy donor's contact only. Using the parameters req_bl_grp and dis_acc(parameters stated above) as weights of a function: φ(req_bl_grp, dis_acc) along with the activation function: **f**(req_bl_grp) we approach the hidden layers finally leading to the output layer. The geographic constraints will act as bias for the input layer. Hence an interested healthy donor's contact details are retrieved from the cloud and shared with the patient in need in a time of crisis (Fig. 6) [8].

Collaborative Filtering Technique
Collaborative filtering [3] is most commonly used in filtering and recommendation mechanisms. Introduced in the 2000's in the industry for the first time this technique is most commonly used now for over a decade now. Collaborative filtering uses algorithms to filter data from user reviews to make personalized recommendations for users with similar preferences. It can filter out items that a user might like on the basis of reactions by similar users. It works by searching a large group of people and finding a smaller set of users with tastes similar to a particular user. This technique is the correct choice for the recipients to find their required donors. The collaborative filtering is applied on

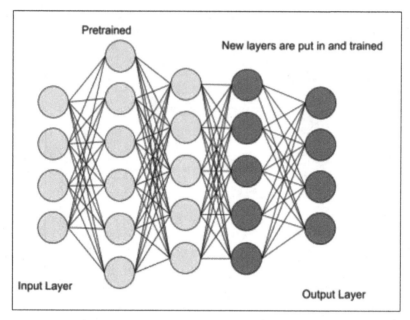

Fig. 5. Transfer Learning Neural Network Architecture.

the healthy donor dataset which was classified in section above this will give optimized results thereby helping FYD to serve more in need (Fig. 7) [2,10].

Working Principle

The application works on collaborative filtering the principle of data storage and retrieval from cloud-based servers are dependent on the Donor to Acceptor Mapping according to which search query placed by a user. A set of healthy donors is classified from the interested donors for a particular blood group upon which the further operations are performed. The search dimensions are tuned based on the category of the flag raised, as discussed above.

Implementation Result and Future Work

Implementation. The proposed solution is implemented in an Android App using the Flutter Framework for cross platform support. For cloud-based data storage and retrieval mechanism Flutter-Firebase is used. Local auth flutter package is used for the authentication part, especially for security authentication. Text to Speech Flutter package is used for voice response. Various other necessary packages can be imported from time to time to improve the user experience and to manage the efficiency of the app.

Result. The android application version of Find Your Donor is now much efficient for secondary level test phase, since primary level testing is successful.

Future Work. Find Your Donor (FYD) will be available both on Android and iOS devices. A dataset with the required attributes is to be implemented for analysis purposes. After successful split of the dataset into train set, dev set and test sets. The training data

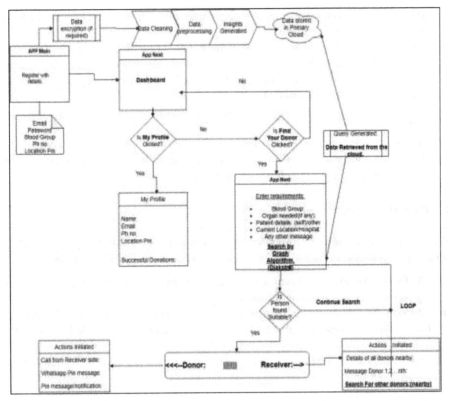

Fig. 6. Flow Diagram: Find Your Donor at a glance.

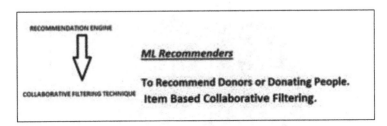

Fig. 7. Recommendation Engine

set now will be used to train the model. This will successfully establish the medium for the next phase of work in app development via which we establish a user interface and a global network among them.

4 Novelty/USP

The following key features determine the USP of Find Your Donor (FYD):

i. At present there is no organization in our country offering services at par with which Find Your Donor is capable of.
ii. If brought into practical use life of many patients can be saved in the long run. Mortality rate can be decreased due to casual accidents and our country can be prepared for any organ transplants or related issues where finding a perfect match is a very tedious task at the present.
iii. This is one of the unique business ideas in the health-care industries, which serves the nation.
iv. After successful financial acquisition in the market, the concept of Find Your Donor application can be extended to finding organ donors like Kidneys, Heart etc., In which finding a correct match still remains a major issue.

5   Revenue Model

Without a basic revenue any proposed idea is about to see a fall. Hence for FYD to function properly the proposed revenue model is as described below in Fig. 8:

i. FYD's is major revenue is going to be driven by charitable-trusts & non-governmental organization NGOs working for the society.
ii. Secondly NGO's individual donations will drive a part of its revenue. Except for blood donation, organ donation and other services should be available for a nominal fee.
iii. Individuals can also be a part of FYD by being in a remaining 30% equity-share model.

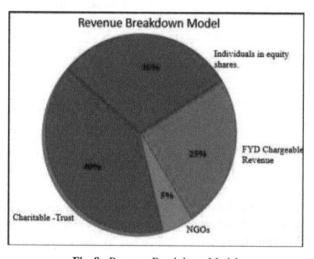

Fig. 8. Revenue Breakdown Model

6   Risk and Mitigation Plan

There is a high risk that fraudsters might scam the patients or their relatives in those chaotic situations, especially the ones having rare blood groups. So, to mitigate this issue, a voice enabled interface is being developed along with a 24 × 7 customer care service associated with Find Your Donor (FYD) which will check and filter out any kind of scams or fraudulence activity and stop them.

7   Conclusion

Find Your Donor App primarily focuses on resolving the problems faced by the people who are in need of blood donors in case of an emergency or contacting nearest hospitals so that a perfect match of organs like kidneys, heart can save the people of all age groups especially the youths who will shape the nation. The proposed app ensures that any donor or user enjoys is able to participate in the digital sphere of India. This app will not only provide an edge over the hassle-free queues over hospitals and phone calls to anonymous persons at late hours but will also motivate the upcoming generations to come up with new ideas to help the society. People who are suffering from thalassemia, organ disorders or any other rare diseases can now ease out a bit of their hardships by introducing such an App for their daily health care needs.

References

1. Bharucha, J.: Tackling the challenges of reducing and managing food waste in Mumbai restaurants. Brit. Food J. **120**(3), 639–649 (2018)
2. He, X., Liao, L., Zhang, H., Nie, L., Hu, X., Chua, T.S.: Neural collaborative filtering. In: Proceedings of the 26th International Conference on World Wide Web, pp. 173–182 (2017)
3. Huang, Z., Zeng, D., Chen, H.: A comparison of collaborative-filtering recommendation algorithms for e-commerce. IEEE Intell. Syst. **22**(5), 68–78 (2007)
4. Oswalt, R.M.: A review of blood donor motivation and recruitment. Transfusion **17**(2), 123–135 (1977)
5. Newman, B.H., Newman, D.T., Ahmad, R., Roth, A.J.: The effect of whole-blood donor adverse events on blood donor return rates. Transfusion **46**(8), 1374–1379 (2006)
6. Shi, L., et al.: Blood donor management in China. Transfusion Med. Hemother. **41**(4), 273–282 (2014)
7. Abdullah, T.A., Zahid, M.S.M., Ali, W.: A review of interpretable ML in healthcare: taxonomy, applications, challenges, and future directions. Symmetry **13**(12), 2439 (2021)
8. Montavon, G., Samek, W., Müller, K.R.: Methods for interpreting and understanding deep neural networks. Dig. Signal Process. **73**, 1–15 (2018)
9. Miikkulainen, R., et al.: Evolving deep neural networks. In: Artificial Intelligence in the Age of Neural Networks and Brain Computing, pp. 293–312. Academic Press (2019)
10. Cakir, O., Aras, M.E.: A recommendation engine by using association rules. Procedia Soc. Behav. Sci. **62**, 452–456 (2012)
11. Kapucu, N., Arslan, T., Demiroz, F.: Collaborative emergency management and national emergency management network. Disast. Prevent. Manag. Int. J. **19**(4), 452–468 (2010)

Machine Learning, Deep Learning
and Text Analytics

Deep Learning-Based Multiple Detection Techniques of Covid-19 Disease From Chest X-Ray Images Using Advanced Image Processing Methods and Transfer Learning

Arif Hussain[1], Rohini Basak[1] , and Sourav Mandal[2(✉)]

[1] Department of Informtion Technology, Jadavpur University, Kolkata 700098, West Bengal, India
[2] School of Computer Scinece and Engineering, XIM University, Bhubaneswar 752050, Odisha, India
sourav@xim.edu.in

Abstract. The 2019 new coronavirus (COVID-19), which originally surfaced in the Chinese city of Wuhan in December 2019, expanded quickly, and caused a pandemic. It has created a terrible influence on daily lives, public health, and the global economy. To stop the pandemic from spreading further and to treat the affected individuals as soon as feasible, it is essential to identify positive cases as soon as possible. One of the primary issues in the present COVID-19 pandemic is the early detection and diagnosis of COVID-19, as well as the precise separation of non-COVID-19 cases at the lowest cost and in the early stages of the disease. This study compares the usage of the most recent Convolutional Neural Network (CNN), a deep learning methodology employing networks, with a transfer learning technique using VGG and ResNet. With their assistance, we created models that can recognize COVID-19 in X-Ray photos. We have applied techniques of image preprocessing using Contrast Limited Adaptive Histogram Equalization (CLAHE) and Adaptive Histogram Equalization (AHE). We built the model by applying CNN. We have also applied transfer learning techniques using Visual Geometry Group (VGG16) and Residual Network (ResNet50) after image preprocessing using CLAHE and AHE. We have compared the accuracy of all models. The CLAHE technique along with ResNet50 works well as we achieve an accuracy of 0.99 It can be stated that ResNet50 with CLAHE played a very significant role in determining the covid-19 in X-ray images.

Keywords: CNN · CLAHE · AHE · COVID-19 · VGG16 · ResNet50

1 Introduction

Coronavirus disease 2019 (COVID-19) is an infectious illness brought on by the coronavirus strain known as severe acute respiratory syndrome coronavirus 2 (SARS-CoV-2). The World Health Organization (WHO) declared the illness a pandemic on March 11, 2020. According to Worldometer, a database for tracking Covid cases worldwide, as of

July 2022, there were over 571 million cases registered globally, with a fatality rate of 1.1% of all closed cases. Non-therapeutic procedures can lessen the enormous strain on healthcare systems while offering the most accurate and reliable diagnostic techniques for COVID-19. Because it is widely accessible. Artificial intelligence (AI) and medical pictures have been discovered to be helpful for quick diagnosis and treatment of COVID-19-infected patients. Thus, it has become urgently necessary to design and implement AI techniques for COVID-19 image classification in a short amount of time with minimal data in order to combat the present pandemic. Others have proven that medical imaging plays a crucial role in enabling a quick diagnosis of COVID-19 [1], and the combination of AI and chest imaging can assist explain the consequences of COVID-19 [2].

Regarding the image analysis of COVID-19, chest X-rays are an imaging technique that hospitals have embraced to diagnose COVID-19 infection, especially the first image-based method employed in Spain [3]. According to the protocol, if a patient's examination reveals a clinical suspicion of infection, a sample of nasopharyngeal exudate is taken for a reverse-transcription polymerase chain reaction (RT-PCR) test, and a chest X-ray is then taken. Since it can take several hours for the PCR test results to be made available. With the prevalence of chest radiology imaging systems in healthcare systems, radiography in medical examinations can be quickly performed and is now widely accessible. However, radiologists' ability to interpret radiography images is constrained by their inability to recognize the fine visual details present in the images. Since CNN may uncover patterns in chest X-rays that ordinarily would not be identified by radiologists [4], there have been numerous studies mentioned in the next Sect. 2 literature survey regarding different deep learning models for recognizing COVID-19.

The study suggested using CLAHE and AHE in combination with intelligent deep learning architectures such CNN architecture, VGG16, and ResNet50 to detect COVID-19 disease. Totaling 2295 x-ray pictures for the investigation, the dataset includes 712 COVID 19-infected photos and 1583 uninfected images from publicly accessible GitHub and Kaggle. The proposed model was assisted by image processing techniques like enhancement, normalization, and data augmentation to not only avoid over fitting but also to display the best accuracy.

2   Related Works

The COVID-19 was foreseen through machine learning. These forecasts can assist healthcare administrators and policy makers in efficiently planning, allocating, and deploying healthcare resources. An alignment-free method based on machine learning was used in [5] to identify an intrinsic COVID-19 genomic signature. With greater than 90% accuracy, the dataset's 61:8 million bp worth of more than 5000 distinct genomic sequences were evaluated.

A machine learning-based technique was created for a real-time forecast of the COVID-19 epidemic utilizing news alerts from Media Cloud and official health reports from the Chinese Center Disease for Control and Prevention, according to [6].

Using US data beginning with the first case on January 20, 2020, [7] proposed a data-driven ML strategy for the investigation of the COVID-19 pandemic from its early infection dynamics, notably inflation counter through time. [8] used an epidemiological labelled dataset to apply machine learning techniques such as decision trees, logistic regression, naive Bayes, support vector machines, and artificial neural networks. With an accuracy rate of 94.99%, the decision tree model ended up being the most accurate of all models created.

Data mining and a deep learning pilot study were used in the work of [9] to estimate the occurrence of COVID-19 in Iran using Google trend data. To estimate the number of COVID-19 positive cases, linear regression models and long short-term memory were both applied. The models were assessed using 10 folds of cross-validation and the root mean square error (RMSE) measure, respectively. The RMSE of the linear regression and long short-term memory models were 27.187 and 7.562, respectively. In the work of [10], a novel CNN model with end-to-end training was suggested. The experiments for the study used a dataset with 180 COVID-19 and 200 normal (healthy) chest X-ray images. As a result of the study, a classification accuracy performance measurement of 91.6% was attained. Individuals with COVID-19 symptoms were found in [11] from 400 chest X-ray pictures using eight different deep learning techniques: VGG16, InceptionResNetV2, ResNet50, DenseNet201, VGG19, MobilenetV2, NasNetMobile, and ResNet15V2.In chest X-ray datasets, NasNetMobile outperformed all other models, with an accuracy of 93.94%.

The 1428 chest radiography of patients with common bacterial pneumonia, verified COVID-19 positive, and healthy individuals are part of the dataset examined in [12] (no infection). The pre-trained VGG16 model was used in this study to successfully train the network on comparatively small chest radiography for classification tasks. For two classes (COVID and non-COVID) and three output classes (COVID, non-COVID pneumonia, and normal), the study showed accuracy rates of 96% and 92.5%, respectively. In the study [13], transfer learning was carried out using the VGG16 CNN and Resnet50, both of which were trained using color camera pictures from ImageNet.10-fold cross-validation was carried out to achieve an overall accuracy of 89.2% in order to evaluate the viability of using chest X-rays to diagnose COVID-19. [14] shows that deep CNNs could accurately and effectively discriminate 21,152 normal and abnormal chest radiographs based on their successful performance in diagnosis accuracy. For the classification of normal versus pneumonia, the CNN model had an accuracy of 94.64%, a sensitivity of 96.5%, and a specificity of 92.86% after being pre-trained on datasets of elderly patients and fine-tuned on pediatric patients. In [15], the COVID-Net architecture was suggested. The model was trained using an open resource called COVIDx, which includes 13,975 X-ray photos, even though only 266 of 358 cases were categorized as COVID-19. They achieved accuracy was 93.3%. Decompose, Transfer, and Compose (DeTraC), a method introduced by [16] was suggested for using chest X-ray images to diagnose COVID-19. An extensive global hospital picture dataset is used in the investigation. DeTraC can identify COVID-19 X-ray images from healthy people and patients with severe acute respiratory syndrome with an accuracy of 93.1%, addressing any irregularities in the image dataset by investigating its class borders.

[17] compared the accuracy of the Inception V3, Xception, and ResNeXt models. 6432 chest x-ray scan samples from the Kaggle repository were gathered in order to analyse the model's performance, of which 5467 were used for training and 965 for validation. As compared to other models, the Xception model has the highest accuracy (i.e.,

97.97%) for detecting chest X-ray pictures. [18] assessed the effectiveness of the most recent pre-trained model ResNet-50 on the 1000-sample COVID-Chest X-ray dataset. 96% accuracy was attained by Resnet-50 with 0.98 sensitivity and 0.95 specificity.

[19] classified the COVID-19 CXR images using a brand-new deep learning model dubbed attention-based VGG-16, which builds on top of VGG-16. We used three COVID-19 CXR datasets to assess our methodology. The evaluation's accuracy for dataset 3 was 87.49%. The VGG16 with transfer learning utilizing x-ray pictures was built in the work of [27], and 96% accuracy was attained. In the work of [28], ResNet201 was built using x-ray pictures, and 84% accuracy was obtained for COVID19 detection. The use of contemporary deep learning models (VGG16, VGG19, DenseNet121, Inception-ResNet-V2, InceptionV3, Resnet50, and Xception) in a comparative research to deal with the identification and classification of coronavirus pneumonia from pneumonia patients is done in the work of [20]. In this investigation, DenseNet121 performed better than the other models, with a 99.48% accuracy rate.

3 Materials and Methods

3.1 Tools

- **OpenCV-** OpenCV is a sizable open-source library for image processing, machine learning, and computer vision, and it currently plays a significant part in real-time operation, which is crucial in modern systems. With it, one may analyze pictures and movies to find faces, objects, and even human handwriting. Python may process the OpenCV array structure during analysis when it is integrated with different libraries, such as NumPy. We use vector space and apply mathematical operations to these aspects to identify the image pattern [20] and its numerous features.
- **Keras:** An open-source, end-to-end machine learning platform is TensorFlow [21]. Researchers can advance the state-of-the-art in ML thanks to its extensive, adaptable ecosystem of tools, libraries, and community resources, while developers can simply create and deploy ML-powered apps. The model is trained using the Keras API. It is a high-level neural network API wi1th the express purpose of making the deep learning model construction and training simple. It is a Python open-source library that utilizes Tensorflow [21].
- **VGG 16**: A convolutional neural network with 16 layers is called VGG-16. ImageNet database contains a pre-trained version of the network that has been trained on more than a million images [22].
- **ResNet50:** A convolutional neural network with 50 layers is called ResNet-50. ImageNet database contains a pre-trained version of the network that has been trained on more than a million images [22].

3.2 The Datasets

To test this investigation, a database of chest X-rays [23] was used. There are 1583 normal X-ray images and 712 COVID 19-infected images totaling 2295 in the database. Techniques for enhancing data and images are used to increase the quantity and variety of images sent to the classifier for classification. For all the retrieved data from the original dataset, horizontal flip, rotation, width shift, and height shift were applied as image augmentations. Vertical flip was not used since X-ray images of the chest are not vertically balanced. Table 1 displays each augmentation parameter. The dataset was expanded to include 1257 COVID 19 infected images and 2849 healthy X-ray images after image augmentation. Also, the AHE and CLAHE image enhancement algorithms were utilized to normalize images and emphasize tiny characteristics so that machine learning classifiers would notice them. Table 2 describes the total number of training, validation, and test data after augmentation.

Table 1. Parameters of data augmentation.

Augmentation Technique	Range
Horizontal flip	**True**
Rotation range	**−20° to 20°**
Width shift range	**0.0–0.2**
Height shift range	**0.0–0.2**
Zoom range	**0.0–0.2**
Shear range	**0.0–0.2**
Vertical flip	**False**

Table 2. Distribution of data in the dataset

Data set	Count
Training	2898
Validation	724
Testing	424

3.3 Performance Parameters

Following parameters have been used for performance analysis:

- **Accuracy:** It is a performance metric that calculates the proportion of accurate forecasts to all observations.

$$\text{Accuracy} = \frac{\text{True Positive} + \text{True Negative}}{\text{True Positive} + \text{False Positive} + \text{False Negative} + \text{False Positive}} \quad (1)$$

- **Precision:** It is the proportion of observations successfully anticipated as positive to all observations correctly forecasted as positive.

$$\text{Precision} = \frac{\text{True Positive}}{\text{True Positive} + \text{False Positive}} \quad (2)$$

- **Recall:** It is also known as sensitivity because it concentrates on the overall quantity of genuine hits discovered.

$$\text{Recall} = \frac{\text{True Positive}}{\text{True Positive} + \text{False Positive}} \tag{3}$$

- **F1 score:** The weighted average of recall and precision is used. Both false positives and false negatives are considered while calculating this score.

$$\text{F1 score} = \frac{2 * \text{Precision} * \text{Recall}}{\text{Precision} + \text{Recall}} \tag{4}$$

3.4 Deep Neural Architecture

In this section we will analyze the architectures of our model for COVID-19 prediction.

1. Architecture 1: Image Augmentation + CNN
The first architecture, seen in Fig. 1, uses an image as its input and outputs the goal level. Hence, we begin by extracting features from the input image using convolution, and then we use ReLU. The input image is sent into the Conv 2D block after being enhanced. The rectifier function is being used to boost the nonlinearity of our pictures, which will then be followed by the maxpooling layer and dropout regularity.

The convolution is again repeated, but this time there are more filters present.

Then, we use a flatten layer to combine them into a 1D array so we can feed them into a fully linked layer. We employ a thick layer with sigmoid activation function as our final step.

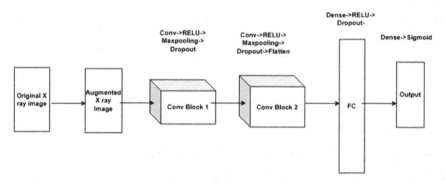

Fig. 1. Architecture 1

2. Architecture 2: Image Augmentation + AHE + CNN
After image augmentation in the second architecture, which is depicted in Fig. 2, we used image enhancement AHE [24]. A digital image processing method called adaptive histogram equalization (AHE) is used to improve the contrast of images.

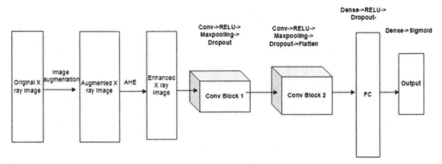

Fig. 2. Architecture 2

3. Architecture 3: Image Augmentation + CLAHE + CNN

After image augmentation in the third architecture, depicted in Fig. 3, we used image amplification Contrast Limited Adaptive Histogram Equalization (CLAHE) [25] to equalize the pictures. Convolution is then applied to the input image to conduct feature extraction, after which the image is sent into the Conv 2D block.

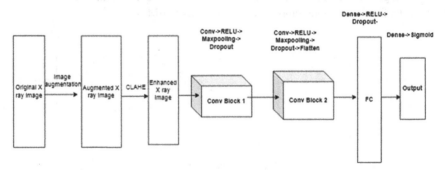

Fig. 3. Architecture 3

4. Architecture 4: Original Images + VGG16

We used advance CNN for transfer learning in the fourth architecture, depicted in Fig. 4, along with VGG16 and the original data set. We removed the last layer of the pre-trained model and replaced it with a linear classifier. Pre-trained weight VGG was employed. We developed a VGG model that pre-trained weights automatically.

The data's input shape has been set to [150, 150]. With include top set to false, we used an ImageNet with pre-trained weights. Then, for the multiclass classifier, we added a flatten layer and employed activation as SoftMax.

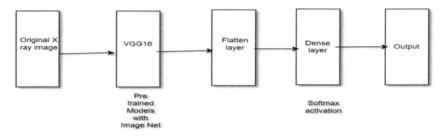

Fig. 4. Architecture 4

5. Architecture 5: VGG16 with AHE

After image augmentation, we used AHE [26] to apply image enhancement in the fifth architecture, which is shown in Fig. 5. Photographic contrast can be improved using the digital image processing technique known as adaptive histogram equalization (AHE). Next, CNN is used, and VGG16 is used to implement transfer learning. In pre-trained weight VGG, we have severed the head. The pre-trained weights and input shape of the data supplied as [150, 150] are automatically loaded into a VGG model. We utilized pre-trained ImageNet with include top set to false. In order to create a multi-class classifier, we added a flatten layer and employed activation as SoftMax.

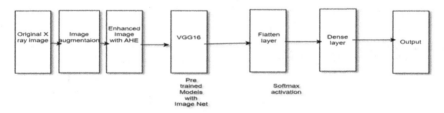

Fig. 5. Architecture 5

6. Architecture 6: VGG16 with CLAHE

After applying image augmentation in the sixth architecture, depicted in Fig. 6, we have applied image enhancement techniques, CLAHE [24] to equalize image. Contrast Limit Adaptive histogram equalization (CLAHE) is a variant of adaptive histogram equalization that limits contrast amplification to reduce noise amplification.

Next, Advance CNN is used, and VGG16 is used to implement transfer learning. In pre-trained weight VGG, we have severed the head. The pre-trained weights and input shape of the data supplied as [150, 150] are automatically loaded into a VGG model.

We utilized pre-trained ImageNet with include top set to false. Lastly, we utilized SoftMax's multi class classifier with flatten layer and activation.

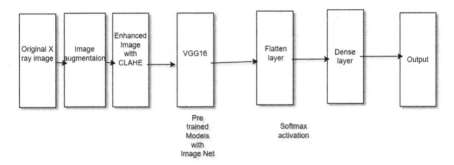

Fig. 6. Architecture 6

7. Architecture 7: Original Images + ResNet50

Using ResNet50 and the original data set, we used advanced CNN in the seventh architecture, which is seen in Fig. 7. We removed the last layer of the pre-trained model and replaced it with a linear classifier. ResNet50 is used with pre-trained weight. We developed a ResNet model that loads the pre-trained weights automatically and specifies the data's input shape as [150, 150]. With include top set to false, we used an ImageNet with pre-trained weights. As SoftMax is employed as a multi-class classifier, a flatten layer is added and activation is performed.

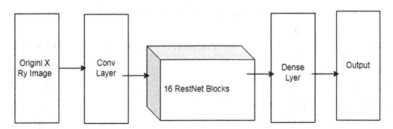

Fig. 7. Architecture 7

8. Architecture 8: ResNet50 with AHE

After image augmentation, we performed image enhancement AHE [24] to the eighth architecture (Fig. 8).

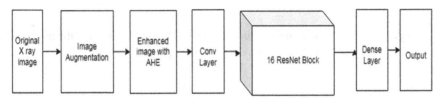

Fig. 8. Architecture 8

9. Architecture 9: ResNet50 with CLAHE (Proposed)

After applying image augmentation in the ninth architecture, depicted in Fig. 9, we have applied image enhancement CLAHE [24] to equalize the image.

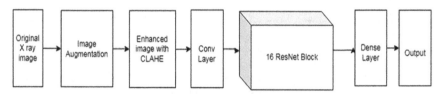

Fig. 9. Architecture 9

4 Experiments and Results

Following the completion of tests using various architectures, we noticed the following performance of the architectures throughout the training phase (Tables 3) and testing phase (Table 4).

Table 3. Performance of the architectures during training phase

Architectures	Accuracy	Precision	Recall	F1 Score
Original image	0.87	0.85	0.90	0.86
Architecture 1	0.88	0.86	0.91	0.87
Architecture 2	0.83	0.82	0.88	0.82
Architecture 3	0.92	0.89	0.93	0.91
Architecture 4	0.99	0.98	0.99	0.98
Architecture 5	0.98	0.98	0.97	0.98
Architecture 6	0.99	0.98	0.99	0.98
Architecture 7	0.99	0.99	0.99	0.99
Architecture 8	0.99	0.99	0.99	0.99
Architecture 9	**0.99**	**0.99**	**0.99**	**0.99**

The graphs of the various architectures depicting the performance of the models on the training and validation datasets are shown below in Figs. 10 through 27:

Table 4. Performance of the architectures during testing phase

Architectures	Accuracy	Precision	Recall	F1 Score
Original image	0.94	0.93	0.95	0.94
Architecture 1	0.96	0.96	0.96	0.96
Architecture 2	0.94	0.93	0.95	0.94
Architecture 3	0.96	0.97	0.95	0.96
Architecture 4	0.98	0.98	0.98	0.98
Architecture 5	0.97	0.98	0.97	0.97
Architecture 6	0.99	0.98	0.99	0.98
Architecture 7	0.98	0.99	**0.99**	0.99
Architecture 8	0.98	0.97	0.98	0.97
Architecture 9	**0.99**	**0.99**	0.98	**0.99**

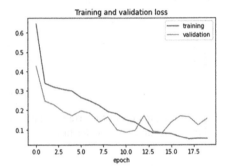

Fig. 10. Graph representing Training and Validation loss of Architecture 1

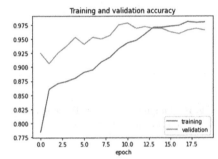

Fig. 11. Graph representing Training and Validation accuracy of Architecture 1

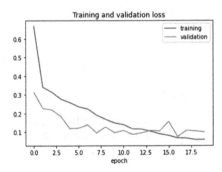

Fig. 12. Graph representing Training and Validation loss of Architecture 2

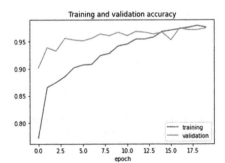

Fig. 13. Graph representing Training and Validation accuracy of Architecture 2

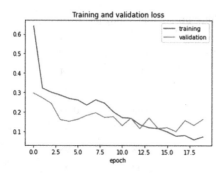

Fig. 14. Graph representing Training and Validation loss of Architecture 3

Fig. 15. Graph representing Training and Validation accuracy of Architecture 3

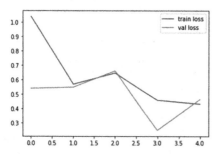

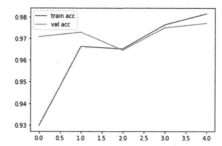

Fig. 16. Graph representing Training and Validation loss of Architecture 4

Fig. 17. Graph representing Training and Validation accuracy of Architecture 4

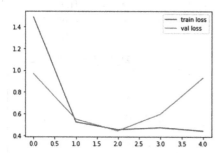

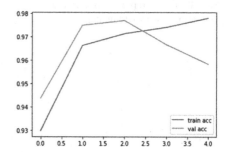

Fig. 18. Graph representing Training and Validation loss of Architecture 5

Fig. 19. Graph representing Training and Validation accuracy of Architecture 5

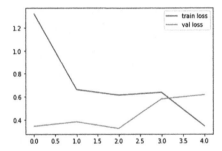

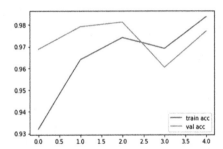

Fig. 20. Graph representing Training and Validation loss of Architecture 6

Fig. 21. Graph representing Training and Validation accuracy of Architecture 6

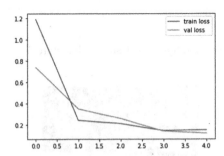

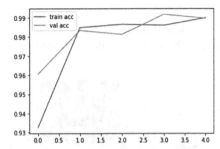

Fig. 22. Graph representing Training and Validation loss of Architecture 7

Fig. 23. Graph representing Training and Validation accuracy of Architecture 7

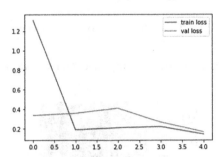

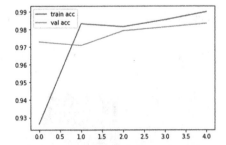

Fig. 24. Graph representing Training and Validation loss of Architecture 8

Fig. 25. Graph representing Training and Validation accuracy of Architecture 8

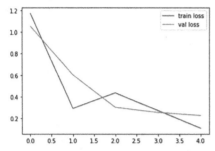 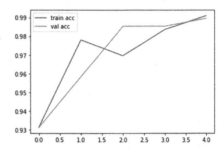

Fig. 26. Graph representing Training and Validation loss of Architecture 9

Fig. 27. Graph representing Training and Validation accuracy of Architecture 9

Confusion Matrices
Figure 28 and Fig. 29 show confusion matrix of Architecture 1 and Architecture 2.

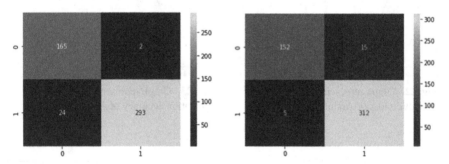

Fig. 28. Confusion matrix of Architecture 1   **Fig. 29.** Confusion matrix of Architecture 2

Figure 30 and Fig. 31 show confusion matrix of Architecture 3 and Architecture 4.

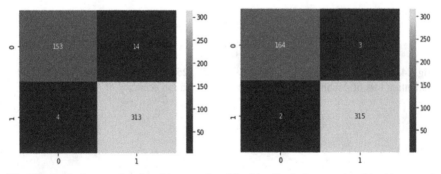

Fig. 30. Confusion matrix of Architecture 3   **Fig. 31.** Confusion matrix of Architecture 4

Figure 32 and Fig. 33 show confusion matrix of Architecture 5 and Architecture 6.

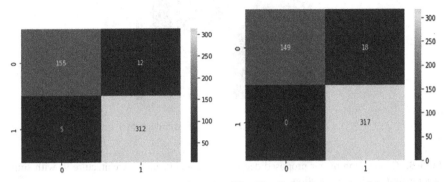

Fig. 32. Confusion matrix of Architecture 5   **Fig. 33.** Confusion matrix of Architecture 6

Figure 34 and Fig. 35 show confusion matrix of Architecture 7 and Architecture 8.

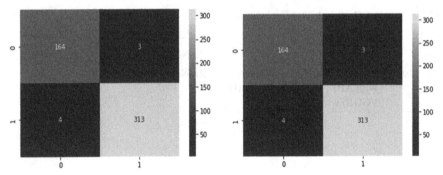

Fig. 34. Confusion matrix of Architecture 7   **Fig. 35.** Confusion matrix of Architecture 8

Figure 36 shows confusion matrix of Architecture 9. Table 5 shows how well our proposed system (Architecture 9) performed on the same dataset when compared to comparable contemporary systems. Our system performs noticeably better than competing systems by a wide margin.

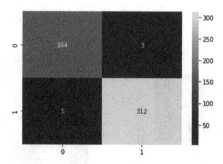

Fig. 36. Confusion matrix of Architecture 9

Table 5. Comparison of our proposed deep learning architecture (Architecture 9) with similar systems for COVID-19 detection.

Reference	Architectures	Best Performing architectures	Accuracy
[17]	Xception net, Inception net V3, Inception net V3	Xception net	0.98
[27]	VGG-19 with Transfer Learning	VGG-19 with Transfer Learning	0.96
[18]	ResNet50	ResNet50	0.96
[28]	ResNet201	ResNet201	0.84
Our Model	ResNet50 with CLAHE, ResNet50 With AHE, VGG16 with CLAHE, VGG16 with HE	**ResNet50 With CLAHE**	0.99

5   Conclusion

This study discusses each of the nine architectural designs, and we discovered that when we used ResNet50 for transfer learning, training data accuracy was 0.99 when used with CLAHE. We also discovered that testing accuracy was 0.99 when used with CLAHE. So, we can draw the conclusion that applying CLAHE along with ResNet50 can significantly improve the likelihood of a patient being diagnosed with the condition with more precision and speed. In subsequent work, we will experiment with high-speed processors, gather information from the COVID-19 omicron and delta variants, and conduct experiments with a larger variety of X-ray image datasets. Additionally, we can contrast by using various transfer learning strategies, such as Xception and Inception.

References

1. Yang, W., et al.: The role of imaging in 2019 novel coronavirus pneumonia (COVID-19). Eur. Radiol. **30**, 4874–4882 (2020)

2. Kundu, S., Elhalawani, H., Gichoya, J.W., Kahn Jr., C.E.: How might AI and chest imaging help unravel COVID-19's mysteries? Radiol. Artif. Intell. **2**(3), e200053 (2020)
3. Gracia, M.M.: Imaging the coronavirus disease COVID-19. [interv.] Mélisande Rouger. s.l. https://healthcare-in-europe.com/en/news/imaging-the-coronavirus-disease-covid-19.html
4. Wong, H.Y.F., et al.: Frequency and distribution of chest radiographic findings in patients positive for COVID-19. Radiology **296**(2), E72–E78 (2020)
5. Randhawa, G.S., Soltysiak, M.P., El Roz, H., de Souza, C.P., Hill, K.A., Kari, L.: Machine learning using intrinsic genomic signatures for rapid classification of novel pathogens: COVID-19 case study. PLoS ONE **15**(4), e0232391 (2020)
6. Liu, D., et al.: A machine learning methodology for real-time forecasting of the 2019–2020 COVID-19 outbreak using Internet searches, news alerts, and estimates from mechanistic models. arXiv preprint arXiv:2004.04019 (2020)
7. Magdon-Ismail, M.: Machine learning the phenomenology of COVID-19 from early infection dynamics. arXiv preprint arXiv:2003.07602 (2020)
8. Muhammad, L.J., Algehyne, E.A., Usman, S.S., Ahmad, A., Chakraborty, C., Mohammed, I.A.: Supervised machine learning models for prediction of COVID-19 infection using epidemiology dataset. SN Comput. Sci. **2**, 1–13 (2021)
9. Ayyoubzadeh, S.M., Ayyoubzadeh, S.M., Zahedi, H., Ahmadi, M., Kalhori, S.R.N.: Predicting COVID-19 incidence through analysis of google trends data in Iran: data mining and deep learning pilot study. JMIR Public Health Surveill. **6**(2), e18828 (2020)
10. Ismael, A.M., Şengür, A.: Deep learning approaches for COVID-19 detection based on chest X-ray images. Expert Syst. Appl. **164**, 114054 (2021)
11. Ahsan, M.M., Gupta, K.D., Islam, M.M., Sen, S., Rahman, M.L., Shakhawat Hossain, M.: Covid-19 symptoms detection based on nasnetmobile with explainable AI using various imaging modalities. Mach. Learn. Knowl. Extr. **2**(4), 490–504 (2020)
12. Pandit, M.K., Banday, S.A., Naaz, R., Chishti, M.A.: Automatic detection of COVID-19 from chest radiographs using deep learning. Radiography **27**(2), 483–489 (2021)
13. Narin, A., Kaya, C., Pamuk, Z.: Automatic detection of coronavirus disease (Covid-19) using X-ray images and deep convolutional neural networks. Pattern Anal. Appl. **24**, 1207–1220 (2021)
14. Tang, Y.X., et al.: Automated abnormality classification of chest radiographs using deep convolutional neural networks. NPJ Digit. Med. **3**(1), 70 (2020)
15. Wang, L., Lin, Z.Q., Wong, A.: COVID-Net: a tailored deep convolutional neural network design for detection of Covid-19 cases from chest X-ray images. Sci. Rep. **10**(1), 1–12 (2020)
16. Abbas, A., Abdelsamea, M.M., Gaber, M.M.: Classification of COVID-19 in chest X-ray images using DeTraC deep convolutional neural network. Appl. Intell. **51**, 854–864 (2021)
17. Jain, R., Gupta, M., Taneja, S., Hemanth, D.J.: Deep learning based detection and analysis of COVID-19 on chest X-ray images. Appl. Intell. **51**, 1690–1700 (2021)
18. Sreejith, V., George, T.: Detection of COVID-19 from chest X-rays using ResNet-50. J. Phys. Conf. Ser. **1937**(1), 012002 (2021)
19. Sitaula, C., Hossain, M.B.: Attention-based VGG-16 model for COVID-19 chest X-ray image classification. Appl. Intell. **51**, 2850–2863 (2021)
20. Shazia, A., Xuan, T.Z., Chuah, J.H., Usman, J., Qian, P., Lai, K.W.: A comparative study of multiple neural network for detection of COVID-19 on chest X-ray. EURASIP J. Adv. Signal Process. **2021**(1), 1–16 (2021)
21. Faujdar, N., Sinha, A.: Disease detection platform using image processing through OpenCV. In: Computational Analysis and Deep Learning for Medical Care: Principles, Methods, and Applications, pp. 181–215 (2021)
22. Team, SuperDataScience.superdatascience, August 2018. https://www.superdatascience.com/blogs/the-ultimate-guide-to-convolutional-neural-networks-cnn

23. ImageNet. ImageNet. http://www.image-net.org
24. Gladding, P.A., et al.: A machine learning PROGRAM to identify COVID-19 and other diseases from hematology data. Future Sci. OA **7**(7), FSO733 (2021)
25. Attia, S.J.: Enhancement of chest X-ray images for diagnosis purposes. J. Nat. Sci. Res. **6**(2), 43–46 (2016)
26. Giełczyk, A., Marciniak, A., Tarczewska, M., Lutowski, Z.: Pre-processing methods in chest X-ray image classification. PLoS ONE **17**(4), e0265949 (2022)
27. Raghavendra, C., Sriram, S.P., Kumar, V.S., Sudheksha, K., Sri, P.B.: Investigation and implementation of convolutional neural networks with transfer learning for detection of Covid-19. J. Phys. Conf. Ser. **2335**(1), 012023. (2022)
28. Li, A.C., Lee, D.T., Misquitta, K.K., Uno, K., Wald, S.: COVID-19 detection from chest radiographs using machine learning and convolutional neural networks. medRxiv, pp. 2020–08 (2020)

Classification Model to Predict the Outcome of an IPL Match

Poulomi Paul[1]([envelope]) [iD], Pratyay Ranjan Datta[2] [iD], and Ashutosh Kar[2] [iD]

[1] Tata Consultancy Services Ltd., London, UK
paulpoulomi@gmail.com
[2] NSHM Knowledge Campus, Kolkata, India
pratyaydata@gmail.com, ashukar1@gmail.com

Abstract. Indian Premiere League or IPL is a very popular form of cricket in India. With the growing interest in the game of cricket, IPL is becoming increasingly popular among the young and old alike. With so many teams participating in IPL, new and young players are getting an opportunity to showcase their talent, play alongside international players and are duly compensated for their talent. With so many regional teams, various players and a vast array of matches, the franchise owners and the fans are very keen and enthusiastic about the winner of these matches. This paper aims to build a Classification Model to predict the outcome of an IPL match using the history data of IPL matches from 2008–2020 and the initial few matches of IPL 2021. Four models have been built using supervised machine learning algorithms like Logistic Regression, Support Vector Machines (SVC), Decision Trees and Random Forest in Python. These models have been built using various features like Season, Venue, Team1, Team2, Toss_winner, Toss_decision, Result_margin etc. The models predict which of the two teams (Team1 or Team2) wins a particular match. Finally, these models are compared for the accuracy of their prediction and the best model is chosen. The best model could match predict the outcome with an accuracy of 73.60%, better than the other models studied in the Literature Review section. This model can be used as a tool to predict which team will win or lose a match, thus enabling teams to modify their strategies, generating money minting options for betting companies etc.

Keywords: Machine Learning · Supervised Learning · IPL Prediction · Python programming · Cricket · Match outcome

1 Introduction

Cricket, the Gentleman's Game originated in England and is now widely popular throughout the world, especially in countries like India, Pakistan, Australia, New Zealand, South Africa, West Indies etc.

The game of Cricket is played between two teams with 11 players on each side. It is played by a bat and ball on an oval field in a rectangular area (pitch). The team captain who wins the toss decides whether to bat first or bowl first. The teams will aim to score more runs than the opposing team (when batting) and to take as many wickets as possible

S. Dhar et al. (Eds.): AGC 2023, CCIS 2008, pp. 83–111, 2024.
https://doi.org/10.1007/978-3-031-50815-8_6

to prevent the opposing team from scoring runs (when fielding). If the team batting first wins the match, they win by 'runs'. If the team batting second wins, they win by the 'number of wickets' left when they achieved the target.

Formats of International Cricket:

- **Test Match** – Traditional form of cricket, played in a 5 day-format with two innings.
- **One Day International** (ODI) – One innings match of 50 overs per team.
- **Twenty20 International** (T20) – The Newest, shortest and fastest form of Cricket. It comprises a 20 overs-format and was started in 2005.

Indian Premier League (IPL):

IPL is a professional T20 League, started by the BCCI in the year 2008. It usually takes place every year in April and May in India. However, due to the Covid-19 outbreak in 2020, IPL 2020 took place in UAE from September to November 2020.

The IPL is a franchise-based competition with eight teams, each representing an Indian city. Franchises acquire players via an annual player auction, trading players in specific trading windows, and signing players who are not picked up in the auction. The majority of signings occur at auction, where players set themselves a base price and join the franchise that ultimately bids the most for them. At the end of each IPL season, the franchise must decide whether to renew each player's contract or release them. Squads must consist of 19–25 players, with a maximum of eight overseas players. A playing XI must not include any more than four overseas players.

IPL Teams:

1. Kolkata Knight Riders
2. Chennai Super Kings
3. Delhi Capitals (formerly known as Delhi Daredevils)
4. Royal Challengers Bangalore
5. Rajasthan Royals
6. Punjab Kings (formerly known as Kings XI Punjab)
7. Sunrisers Hyderabad (formerly known as Deccan Chargers)
8. Mumbai Indians
9. Gujarat Titans (joined in IPL 2022)
10. Lucknow Super Giants (joined in IPL 2022)
11. Pune Warriors (played in IPL 2011, 2012 & 2013)
12. Kochi Tuskers Kerala (played in IPL 2011)
13. Rising Pune Supergiant (played in IPL 2016, 2017 & 2018)
14. Gujarat Lions (played in IPL 2016, 2017 & 2018)

Game Format

The majority of the tournament follows a simple structure:

- Each of the eight teams play each other home and away, with two points earned for a win, one point for a tie or no result, and none for a defeat.
- At the end of that double round-robin process, the bottom four teams in the table are eliminated and the top four teams progress to the playoffs.
- The teams that finished first and second in the league table will play each other in Qualifier 1. The winner of that match will progress to the final, but the loser is not yet eliminated.
- Meanwhile, the teams that finished third and fourth in the league table will play each other in the eliminator.
- The loser of the eliminator is knocked out, while the winner of the eliminator earns the right to play the loser of Qualifier 1 in Qualifier 2.
- The winner of Qualifier 2 plays the winner of Qualifier 1 in the final, where the winner of the tournament is crowned.

The 2021 Indian Premier League (IPL 14) started in April 2021 but due to rising Covid-19 cases, it was suspended in May. At the time of suspension, 31 of the 60 scheduled matches were yet to be played. The pending matches were played in UAE in September and October 2021. This model was built using the match data available from 2008 to 2021. This study is structured into sections: This study is structured into sections: Objectives of the Study, Literature Review, Research Methodology, Findings and Analysis, Validation, Conclusions, Future Scope of Research, Limitations and Bibliography.

2 Objectives of the Study

1. To analyse the data of past matches and predict which team will win a particular match in IPL.
   a. Analyse the data of previous IPL matches i.e., 2008 to 2021.
2. To build a classification model which will identify if a team will win or lose a particular IPL match.
   a. Build multiple classification models and identify which model gives the best prediction.

3 Literature Review

IPL winner prediction has gained popularity post the massive success of its various seasons. Many researchers have tried different techniques for prediction. *Sinha, A.* (2020) elaborates on an IPL match predictor, which is an ML-based prediction approach built using the KNIME Tool with the intelligence of Naive Bayes network and Euler's strength calculation formula. It uses the history data set and stats along with weighted factors: Toss, Home Ground, Captains, Favourite Players, Opposition Battle, Previous Stats etc. to train the model and predict the outcome. The history data used for this research was scrapped from ICC's T20 top 100 players and Cricbuzz website to obtain data for Batsmen, Bowlers and All-Rounders. Different features were taken for these 3 types of cricketers:

- For Batsmen: Innings, Runs Scored, Batting Average, Batting Strike Rate, Fifties, Fours, Sixes.
- For Bowlers: Innings, Wickets, Economy, Bowling Average, Bowling Strike Rate.
- For All-Rounders: Innings as Batsmen, Runs Scored, Batting Average, Batting Strike Rate, Fifties, Fours, Sixes, Innings as Bowler, Wickets, Economy, Bowling Average, Bowling Strike Rate.

Log transformation was done on the obtained data to make the data normal as the existing data was highly right-skewed. Two models were built: SGD-Regressor and KNN-Regressor. The outcome of these models was compared with Scikit-Learn's model.

- **SGD-Regressor:** The training data was used to build the model and find the best number of epochs using Cross-Validation Data. It was found that as the epochs increase, the error decreases. 2000 epochs were used as at this point there was a steady fall in error rate. The test data set was used to obtain the predicted values, which were compared with the actual values to get the accuracy of the model.
- **KNN-Regressor:** The training data was used to build the model and find the best number of neighbours using Cross-Validation Data. It was found that as the value of K increases, the error also increases i.e., larger values of K will lead to underfitting and smaller values of K will lead to overfitting. Therefore, an optimal number of neighbours was found. Once done, the test data set was used to obtain the predicted values, which were compared with the actual values to get the accuracy of the model.

For both models, it was found that Scikit-Learn's model produced almost similar results. The total number of features selected was 17. Out of these, features were chosen selectively and the ML models were built. The model with all 17 features achieved the best accuracy of 90%.

Vistro, Rasheed and David (2019) researched to predict the winner of IPL, taking into account the historical match data of IPL from 2008 to 2017 using Machine Learning Models. The main aim of this research is to predict or classify which of the 13 IPL teams will win. To train the ML models, various common features were considered including batting, bowling, fielding, player performance and team performance along with unique features like key players, pitch condition, home ground advantage and weather conditions. The feature selection was done using data visualization. This analysis of the IPL dataset was done using the SEMMA data mining method. Post data pre-processing and feature selection, three ML models were built and compared for accuracy. The historical dataset was split into training and testing sets for classification models – Decision Tree Classifier, Random Forest Classifier and XGBoost Classifier. These models were built in Python and for each model, parameter tuning was done to get the best model accuracy and confusion matrix analyzed to visualize their performance. It was found that XGBoost Classifier had an accuracy of 94.23% without parameter tuning and Decision Tree Classifier achieved an accuracy of 94.87% post fine-tuning of parameters. However, Random Forest Classifier did not fare as well as these two models.

Kapadia, Abdel-Jaber, Thabtah and Hadi (2020) built Machine learning models to predict the outcome of an IPL match based on certain factors like home ground advantage and toss decision. History data of IPL matches from 2008 to 2017 was taken from Kaggle to build the ML models. Four models were built - Naïve Bayes, Random Forest, K-nearest neighbour and Decision Tree. The dataset was divided into two segments: IPL winner based on the impact of home ground advantage and that of toss decision. The dataset with the impact of home ground advantage had 6 predictor variables and the one with the effect of toss decision had 7 predictor variables. These features were selected using Correlation-based Feature Selection, Information Gain (IG), Relief and Wrapper. These two datasets were given as separate inputs to the above-mentioned ML models and were tested using ten-fold cross-validation with stratification in addition to accuracy, precision, recall and confusion metrics.

For the 1st case: Impact of home ground advantage to match result, it was found that the best accuracy was achieved by Naïve Bayes (57%) followed by Random Forest, decision tree and at last KNN model. For the 2nd case: The effect of the toss decision on the match result, the result of the match was based on the toss winning team. Here the best accuracy was achieved by the KNN model (62%) followed by Naïve Bayes. Therefore, it can be concluded that the model built using Effect of toss decision has generated better prediction of IPL match winner as compared to the impact of home ground advantage.

Patel and Pandya (2019) focused their research on the fact that IPL matches give opportunities to various new players and this paper mainly researches on their performance based on which decision will be made whether to pick or not to pick them. The aim of this research is to build a model that will help to decide whether to pick a player or not in fantasy cricket team like Dream11. The history ball-by-ball data of IPL 2008–2018 has been used to train the supervised machine learning techniques. The data has been split into two sections (i) Batsman: if the batsman has scored greater than 40 runs but less than 80 runs, then pick else don't pick. If the batsman has scored more than 80 runs, then pick the player and consider him for the position of captain. (ii) Bowler: if the bowler has taken 2 or 3 wickets, then pick else don't pick. If the bowler has taken more than 3 wickets then consider him for the position of captain. By classifying which players to pick and not to pick, this problem is converted into a classification problem. Four machine learning models were built and compared for accuracy. For the purpose of simplicity only one batsman: Virat Kohli and bowler: Lasith Malinga were chosen for this analysis.

The first model was Decision Tree – using this modelling technique, the cross-validation for batsman was found to be 59.12% with a standard deviation of 4.43%. For bowler the cross-validation score was 59.70% with a standard deviation of 7.93%. The second model was Random Forest - using this modelling technique, the cross-validation for batsman was found to be 72.14% with a standard deviation of 2.95%. For bowler the cross-validation score was 62.05% with a standard deviation of 4.49%. Similarly for the third model, XGBoost, the cross-validation score was 70.12% with SD 3.28% for batsman and for bowler the cross-validation score was 70.70% with SD 3.68%. The fourth model, i.e. Stacking gave the best classification results with cross-validation score of 74.33% with SD 3.88% for batsman and cross-validation score of 82.41% with SD 3.94% for bowler.

Pathak and Wadhwa (2016) predicted the outcome of an ODI (One day International) using supervised learning techniques. History data of ODI matches from 2001–2015 were collected from Cricinfo website for building the model and only those records were considered which had a clear winner. The factors considered in this prediction were Toss outcome, home ground advantage, day/night effect, bat first. These factors were selected on the basis of a research paper (A. Bandulasiri, "Predicting the Winner in One Day International Cricket") and were proven to have significant impact on the outcome of an ODI match. Three ML models were built: Naïve Bayes, Random Forest and Support Vector Machines and these models were applied for each and every ODI team. The history data was split into 80:20 ratio for training and testing purposes. The performance of each of these models were compared and evaluated using balanced accuracy (higher value indicates better classification) and kappa statistic (higher value indicates better classification). The results obtained indicated that SVM model performed better than the other two for all countries in predicting the outcome of an ODI match. However for Australia, kappa statistic was 0 for Random Forest and SVM due to unbalanced classes. But Naïve Bayes model did not show any issue because of this imbalance and performed well.

Prakash, Patvardhan and Lakshmi (2016) have researched on a prediction model to predict the outcomes of the IPL 9 matches. The prediction model built here is a combination of three different models which are built using various factors and Support Vector Machine (SVM) approach. The final outcome of a match is decided by taking a majority vote of all the three models. History data till IPL 8 and T20 stats of the players have been considered for this purpose. 10 metrics or indices were calculated for the players using the T20 and IPL 8 data. The relative importance of each feature were calculated using Recursive Feature Elimination (RFE) with Random Forest algorithm. Based on the selected features and weights, composite index for batting and bowling performance were calculated. These indices have been used in the models.

Model 1: This model compares the overall performance of one team with that of the opposition team. For both the teams, the overall feature score was calculated for all the players who will play a particular match (composite features). For every player in the present team, batting and bowling ranks were assigned and their differences with the ranks of the opposition team players were calculated. Using these 22 features (11 for each of the 2 teams) SVM was used to predict the target variable i.e. if first team wins (1) or second team wins (0).

Model 2: This model is similar to Model 1. It performs the same steps as Model 1 and uses the 22 features. Additionally, the player form in IPL 9 matches already played have been considered. Thus the data used to train the model is different from Model 1. Similar to the previous model, SVM is used to predict whether the first team wins the match or the second team wins the match.

Model 3: This model has a different approach. The overall performance of both the teams are calculated. Then the batting and bowling features are calculated for each player of the two teams. Instead of using 11 feature values for each player of both teams, an average value of the initial 10 batting and bowling indices for the 11 players are used. Therefore, the number of features is reduced to 10. Therefore, using these updated feature values, IPL 8 data is used to predict the match outcome of IPL 9 using SVM (0/1).

The majority outcome of these three models has been considered as the final outcome. The accuracy of this approach was 69.9%.

Singh and Kaur (2017) focused on two objectives – (i) Identifying the form of the players using data visualization (ii) Predicting the winner of an IPL match using KNN algorithm. The uniqueness of this approach is that it has used HBase for storing and manipulating the data. The history data of IPL matches (2008–2017) and the players was scraped from Cricbuzz website. Based on the requirement only 10 fields of Player stats was chosen. Addressing the first objective, data visualization was done using Tableau to identify the top batsmen i.e. a mapping of the runs scored vs the number of matches played. Similarly, top bowlers were identified by mapping the number of wickets taken vs the number of matches played. Top all-rounders were found by mapping the number of runs scored and wickets taken with the number of matches played. And the top scoring teams for all the seasons of IPL were identified. This objective was to identify the best players and therefore help in player auctions.

Addressing the second objective, a match prediction model was built using k-Nearest Neighbours algorithm with k = 4. A field called 'strength' was calculated using the IPL match data like number of matches played, batting and bowling average, runs scored, wickets taken, bowling economy etc. The prediction model was build using the predictor variables like Team1, Team2, toss, match venue, strength to predict which of the two teams won the match. The model was trained using 50% of the data set and tested using the rest 50%. This supervised KNN with k = 4 predicted the match winner with an accuracy of 71%. This model was compared with other models like SVM, logistic regression, decision tree and random forest, however KNN had the best accuracy.

4   Research Methodology

The below steps were followed for this analysis:

1. *Define Problem Statement*: Define the project outcomes, the scope of the effort, objectives, identify the data sets that are going to be used.
2. *Data Collection*: Data collection involves gathering the necessary details required for the analysis. It involves the historical or past data from an authorized source over which predictive analysis is to be performed.
3. *Data Cleaning*: Data Cleaning is the process in which data sets are refined. Unnecessary, erroneous and redundant data are removed.
4. *Data Analysis*: It involves the exploration of data. Data is analyzed thoroughly in order to identify some patterns or new outcomes from the data set. Useful information, some patterns or trends are identified.
5. *Build Predictive Model*: In this stage of predictive analysis, various algorithms are used to build predictive models based on the patterns observed. Hypothesis testing can be done using standard statistic models.
6. *Validation*: In this step, the efficiency of the model is checked by performing various tests. Sample input sets are provided to check the validity of the model. The model needs to be evaluated for its accuracy in this stage.

Thus, proceeding with the above-mentioned method:

- **Problem Statement:** At a time two teams will be playing a cricket IPL match. Out of these two teams, either Team 1 will win or Team 2 will win. Therefore, the goal of this study is to predict if Team 1 will win a particular IPL match or not. The problem at hand is a **binary classification**. **Team 1 will win? (Yes/No)**. If Team 1 does not win the match, it implies that Team 2 will win.
- **Data Collection:** Data of the past IPL matches from 2008–2021 was collected from Kaggle. Two different datasets containing match data of 2008–2020 and 2021 were merged for this analysis.
- **Data Cleaning:** Following the collection of the required datasets, the data was cleaned for analysis. Some basic data cleaning done were:
- **4 records** with match **result** as **NA** were **removed**. 1 from 2011, 2 from 2015 and 1 from 2019 IPL seasons.
- Some teams revamped, ownerships changed and therefore they have **new names** – **Delhi Daredevils** was changed to **Delhi Capitals**, **Deccan Chargers** was changed to **Sunrisers Hyderabad** and **Kings XI Punjab** was changed to **Punjab Kings**.
- Additional column of **Home Ground (Y/N)** was added to dataset to identify the matches won by the teams in their home ground based on the Venue of the match.
- **Exploratory Analysis:** Based of IPL Match data of 2008–2021, the below exploratory analysis was done. It helped in identifying the type of features that can be used to predict the outcome.

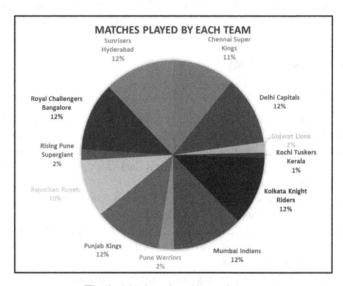

Fig. 1. Matches played by each team.

From Fig. 1, it can be seen that except for Gujarat Lions, Kochi Tuskers Kerala, Pune Warriors and Rising Pune Supergiant, all the other teams have played almost same percentage of matches. There are minor differences for Chennai Super Kings and Rajasthan Royals. It has been mentioned later in per season analysis.

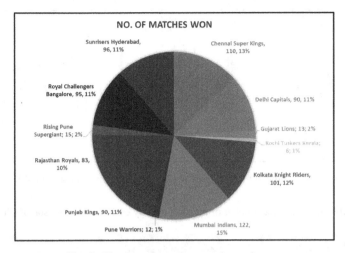

Fig. 2. Number of matches won by each team

From Fig. 2, it can be seen that Mumbai Indians have won the most matches, followed by Chennai Super Kings, Kolkata Knight Riders, Sunrisers Hyderabad and so on.

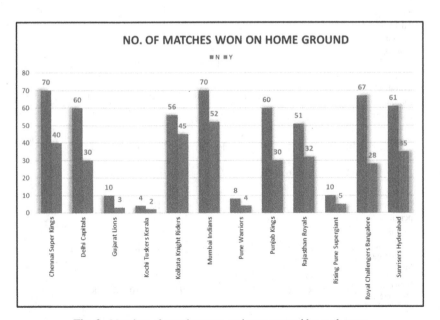

Fig. 3. Number of matches won on home ground by each team

From Fig. 3, it can be seen that although the teams have won quite a few matches on their home ground yet they have won more matches on away ground.

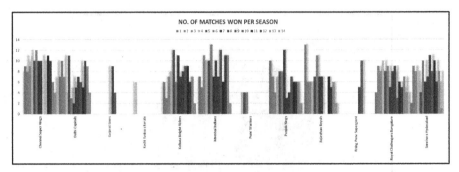

Fig. 4. Shows the number of matches won by each team per season of IPL.

Table 1. Shows the number of matches won by each team per season of IPL.

Teams	Season														
	1	2	3	4	5	6	7	8	9	10	11	12	13	14	To-tal
Chennai Super Kings	9	8	9	11	10	12	10	10			11	10	6	4	110
Delhi Capitals	7	10	7	4	11	3	2	5	7	6	5	10	9	4	90
Gujarat Lions									9	4					13
Kochi Tuskers Kerala				6											6
Kolkata Knight Riders	6	3	7	8	12	6	11	7	8	9	9	6	7	2	101
Mumbai Indians	7	5	11	10	10	13	7	10	7	12	6	11	11	2	122
Pune War-riors				4	4	4									12
Punjab Kings	10	7	4	7	8	8	12	3	4	7	6	6	6	2	90
Rajasthan Royals	13	6	6	6	7	11	7	7			7	5	6	2	83
Rising Pune Su-pergiant									5	10					15
Royal Challeng-ers Banga-lore	4	9	8	10	8	9	5	8	9	3	6	5	7	4	95
Sunrisers Hyderabad	2	9	8	6	4	10	6	7	11	8	10	6	8	1	96
Total	58	57	60	72	74	76	60	57	60	59	60	59	60	21	833

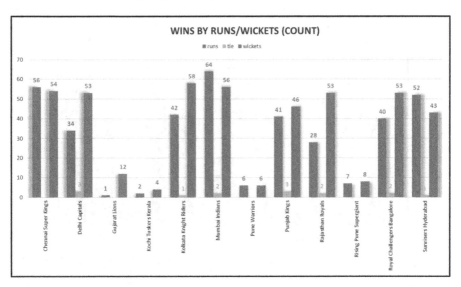

Fig. 5. Wins by runs/wickets of each team

Figure 5 shows the number of wins each team had since Season 1. The wins shown here are segregated by either runs or wickets. Most of the wins are by wickets.

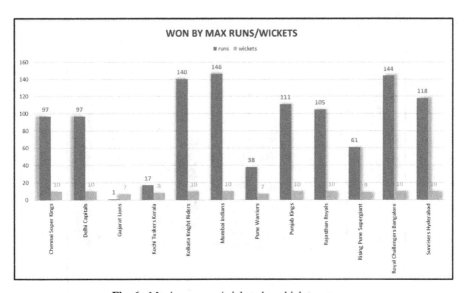

Fig. 6. Maximum runs/wickets by which teams won

Figure 6 shows the maximum gap between the winning team and the losing team (by runs/wickets).

Figure 7 shows the number of matches won by a team by winning the match toss. It can be seen that is most cases teams who have won the toss have also won the match.

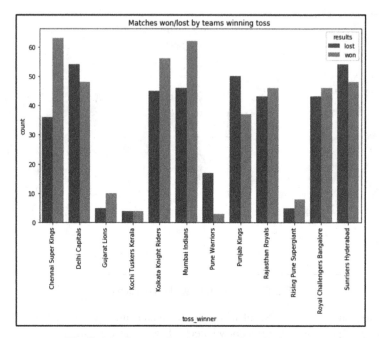

Fig. 7. Matches won/lost by teams after winning toss

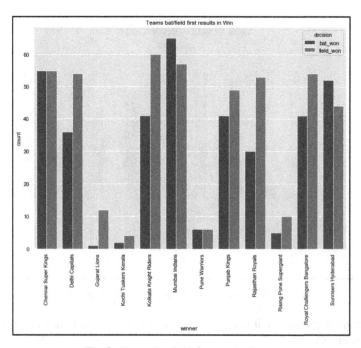

Fig. 8. Teams bat/field first results in win

Figure 8 shows how many times the teams have won if they have batted first or fielded first.

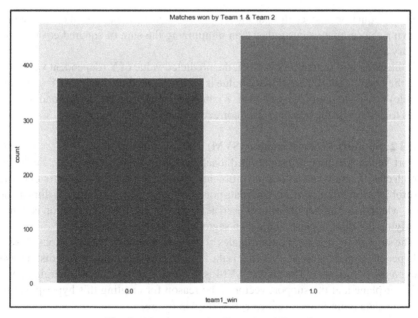

Fig. 9. Matches won by Team 1 and Team 2

Shows the distribution between Team 1 winning the match and Team 2 winning the match (Team 1 losing) Fig. 9. It can be seen that Team 1 has won more as compared to Team 2. But the difference between these two classes is not very high.

- **Build Predictive Model:** Classification is a type of predictive modelling in which a class label is predicted for set of input data. In a classification problem, the **output variable** is a **category**. Based on the input variables, a classification model tries to predict the categorical class label. This is a form of supervised learning.

The predictive modelling of this project has been done in **Python**. Since this is a Classification problem (Binary outcome), **four types of classification models were built and tested for accuracy**. The model with best accuracy was chosen for classification.

Model 1: Logistic Regression
Logistic Regression is a supervised classification algorithm which is used when the dependent variable is discrete and there are several independent variables (continuous or categorical). Example outcome: 0 or 1, true or false, etc. This means the target variable can have only two values, and a sigmoid curve denotes the relation between the target variable and the independent variable.

Logit function is used in Logistic Regression to measure the relationship between the target variable (Y) and independent variables (X). Below equation denotes the logistic regression:

$$Logit\,(p)\; =\; ln(\frac{p}{1-p})\; =\; b_0\; +\; b_1X_1\; +\; b_2X_2\; +\; b_3X_3\; +\; \ldots\; +\; b_kX_k,$$

where, p is the probability of occurrence of the feature.

In the equation above, the parameters are chosen to maximize the likelihood of observing the sample values rather than minimizing the sum of squared errors (like in ordinary regression).

If the estimated probability is >0.5, the predicted value of Y (dependent variable) = 1. On the other hand, if the estimated value is <0.5, the predicted value of Y (dependent variable) = 0. The predicted values of Y can be compared to the corresponding actual values to determine the percentage of correct predictions.

Model 2: Support Vector Machine (SVM)

Support Vector Machine is a supervised machine learning algorithm which can be used for both classification and regression problems. However, it is mostly used for classification problems. In this algorithm, each data point is plotted as a point into an n-dimensional space, where n = number of features used as input. Finally, the classification is done by identifying the right hyper-plane which segregates the two classes well.

The support vectors are the coordinates of the individual input observations closest to the hyper-plane and the SVM classifier is the best decision boundary which distinguishes between the two classes. The goal of SVM is to maximise the distance/margin between the hyper-plane and the support vectors. The reason for creating this hyper-plane is to put a new data point into correct category/class in future.

Model 3: Decision Tree Classifier

Decision Tree is a supervised learning technique that builds classification and regression models in the form of a tree structure. The dataset is broken down into smaller datasets while the tree is incrementally developed.

A decision tree has:

- Internal nodes which represent the features of the dataset (also known as Decision Nodes)
- Branches which represent the decision rules
- Leaf nodes which represent the outcome/classes.

The decision tree is a graphical representation of getting all the possible solutions to a decision based on given conditions. The decisions are taken based on the features of the dataset.

The tree is built from the root node which expands into branches, internal nodes and forms a tree like structure. A decision tree can contain categorical as well as numerical data.

Model 4: Random Forest Classifier

Random Forest is another supervised learning algorithm used for both classification and regression. Random Forest is the most flexible and easy to use algorithm.

Random Forest, as the name suggests comprises a large number of decision trees that operate as an ensemble. The decision trees in the Random Forest are created from randomly selected data samples (of the training dataset). Each tree gives a prediction and the most accurate and stable solution is selected by voting. This algorithm searches for the best feature among a random subset of features while splitting a node, instead of searching the most important feature. This leads to a wide variety that generally results in a better model.

Thus, in random forest, only a random subset of the features is taken into consideration by the algorithm for splitting a node. The trees can be made more random by using random thresholds for each feature rather than searching for the best possible thresholds (like a normal decision tree does).

The above mentioned classification models have been built for IPL match winner prediction in **Python**. The **IPL Dataset** had **833 records** of the matches held so far (from season 1 to season 14).

To build the models, the dataset was randomly split into Training Dataset and Testing Dataset. About **70% of the data** was used as **training data** and the **rest 30%** was used for **testing**.

The models have been validated and tested for accuracy. It is mentioned in the next section.

5 Findings and Analysis

Classification Model was built based on the below Features:

- **Season (1–14)** – Represents the season of a specific match being played. IPL 13 was the last season. IPL 14 started in April 2021. It continued later in September 2021.
- **Venue** – Represents the stadium in which a particular match is being played. It can be a home ground of a team or away ground.
- **Team1** – First Team playing an IPL match.
- **Team2** – Another Team playing against Team1 in that IPL match.
- **Toss_winner** – Team which has won the toss of a match.
- **Toss_decision (bat/field)** – Batting or Fielding decision made by the team which won the toss.
- **Result (runs/wickets)** – Winning factor i.e. winning team won by runs or by wickets.
- **Result_margin** – Winning margin from the losing team.
- **Eliminator (Y/N)** – Was the match being played an eliminating match or not. Y if the type of match is eliminator otherwise N
- **Method (NA or D/L)** – Was the match won by regular method or Duckworth–Lewis–Stern method.

The **Target** Variable: **team1_win (0/1)** – Indicates whether Team1 won the match or not. If Team 1 did not win the match, then Team 2 won the match. 1 indicates Team1 won the match and 0 indicates Team1 lost the match.

Note: The IPL dataset used for this project had a **winner** column with the name of the team that won the match. Using that column, a binary target variable was created i.e., **team1_win**.

The Metrics Used to Analyse the Model

- **Confusion Matrix:** Confusion matrix is a table (Table 2) which describes the performance of a prediction model. A confusion matrix contains the actual values and predicted values.

Table 2. Using this confusion matrix, classification report is generated.

	Predicted 0	Predicted 1
Actual 0	TN/True Negative: when a case was negative and predicted negative i.e. Correct	FP/False Positive: when a case was negative but predicted positive i.e. Incorrect
Actual 1	FN/False Negative: when a case was positive but predicted negative i.e. Incorrect	TP/True Positive: when a case was positive and predicted positive i.e. Correct

- **Classification Report:** The classification report used to measure the quality of predictions from a classification model. Classification Report also displays the Precision, Recall, F1 and Support scores for the model.

  - *Precision* score accuracy of positive predictions. It is calculated as:

$$\text{Precision} = \frac{TP}{(TP + FP)}$$

  - *Recall* is fraction of positives that were correctly identified. It is calculated as:

$$\text{Recall} = \frac{TP}{(TP + FN)}$$

  - *F1 Score* is a measure of a model's accuracy on a dataset, and is defined as the harmonic mean of the model's precision and recall. It is calculated as:

$$\text{F1 score} = 2 * \frac{(Recall * Precision)}{(Recall + Precision)}$$

  - *Support* scores are the amount of data tested for the predictions.

- **Accuracy:** Accuracy score is the percentage of accuracy of the predictions made by the model. It can be calculated as:

$$\text{Accuracy} = \frac{TP + TN}{(TP + FP + TN + FN)}$$

- **Lift Score:** Lift measures the degree to which the predictions of a classification model are better than randomly-generated predictions. Lift Score > 1 indicates that the model can be useful.
- **ROC Curve:** A receiver operating characteristic curve, or ROC curve, is a graphical plot that illustrates the diagnostic ability of a binary classifier system as its discrimination threshold is varied.

The ROC curve shows the trade-off between sensitivity (TPR) and specificity (1 − FPR). As a baseline, a random classifier is expected to give points lying along the diagonal (FPR = TPR). The closer the curve comes to the 45-degree diagonal of the ROC space, the less accurate the test.

6   Model Evaluation and Output Interpretation

The models were evaluated using the Testing data i.e. randomly selected 30% records of the IPL dataset.

Model 1: Logistic Regression

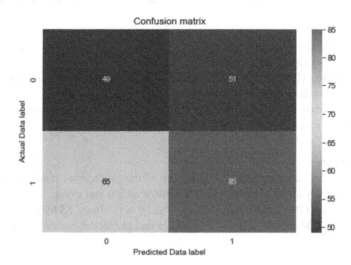

Fig. 10. Logistic Regression Confusion Matrix

From the above <u>Confusion Matrix</u> (Fig. 10), it can be seen that:
Actual Positives = 150, Actual Negatives = 100 and Total = 250. Out of which:

- TP = True Positives = 85
- TN = True Negatives = 49
- FP = False Positives = 51
- FN = False Negatives = 65 i.e.

- Out of all the predictions, the **classifier predicted yes** for the total **136 times**, out of which **85 were actual yes**.
- Out of all the predictions, the **classifier predicted no** for the total **114 times**, out of which **49 were actual no**.

Classification Report

The classification report (Table 3) displays the Precision, Recall, F1 and Support scores for the model.

- The **precision** for Team 1 winning the match is **0.62** and for Team 2 winning (or Team 1 losing) the match is **0.43**.

Table 3. Logistic Regression Classification Report

	precision	recall	f1-score	support
0	0.43	0.49	0.46	100
1	0.62	0.57	0.59	150
accuracy			0.54	250
macro avg	0.53	0.53	0.53	250
weighted avg	0.55	0.54	0.54	250

- **Recall** for Team 1 winning the match is **0.57** and for Team 2 winning (or Team 1 losing) the match is **0.49**.
- The **F1-score** of Team 1 winning the match is **0.59** and for Team 2 winning (or Team 1 losing) the match is **0.46**.
- **Support scores**: In the IPL data-set the data tested for Team 1 winning the match is **150** and for Team 2 winning (or Team 1 losing) the match is **100**.

Accuracy Score

Accuracy score is the percentage of accuracy of the predictions made by the model. Here, the accuracy score of the model is **0.5360**, which is **not good**.

Accuracy = (TP + TN)/Total = (85 + 49)/250 = 0.536 or, **53.60%**

Lift Score: **1.0416**. Lift Score > 1 indicates that the model can be useful.

ROC Curve

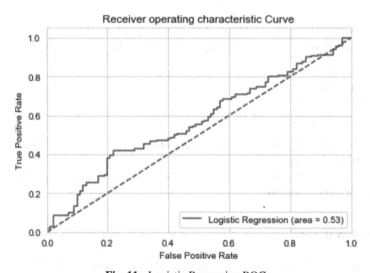

Fig. 11. Logistic Regression ROC

The dotted line represents the ROC curve of a purely random classifier; a good classifier stays as far away from that line as possible (toward the top-left corner). All points above baseline (red dotted line) correspond to the situation where the proportion of correctly classified points belonging to the Positive class is greater than the proportion of incorrectly classified points belonging to the Negative class.

In the above ROC curve (Fig. 11), it can be seen that the **ROC curve is very close** to the baseline. Thus, it can be inferred that this is not a good model for this classification.

Model 2: Support Vector Machine (SVM)

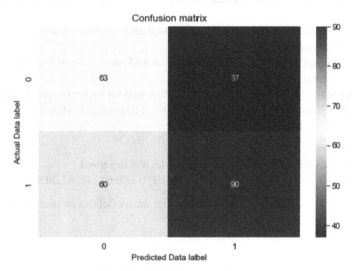

Fig. 12. SVM Confusion Matrix

From the above <u>Confusion Matrix</u> (Fig. 12), it can be seen that:
Actual Positives = 150, Actual Negatives = 100 and Total = 250. Out of which:

- TP = True Positives = 90
- TN = True Negatives = 63
- FP = False Positives = 37
- FN = False Negatives = 60 i.e.

• Out of all the predictions, the **classifier predicted yes** for the **total 127 times**, out of which **90 were actual yes**.
• Out of all the predictions, the **classifier predicted no** for the **total 123 times**, out of which **63 were actual no**.

Classification Report

The classification report (Table 4) displays the Precision, Recall, F1 and Support scores for the model.

• The **precision** for Team 1 winning the match is **0.71** and for Team 2 winning (or Team 1 losing) the match is **0.51**.

Table 4. SVM Classification Report

	precision	recall	f1-score	support
0	0.51	0.63	0.57	100
1	0.71	0.60	0.65	150
accuracy			0.61	250
macro avg	0.61	0.61	0.61	250
weighted avg	0.63	0.61	0.62	250
Accuracy of SVM classifier on test set: 0.6120				

- **Recall** for Team 1 winning the match is **0.60** and for Team 2 winning (or Team 1 losing) the match is **0.63**.
- The **F1-score** of Team 1 winning the match is **0.65** and for Team 2 winning (or Team 1 losing) the match is **0.57**.
- **Support scores**: In the IPL data-set the data tested for Team 1 winning the match is **150** and for Team 2 winning (or Team 1 losing) the match is **100**.

Accuracy Score

The accuracy score of the model is **0. 6120**, which is **not good**.

Accuracy = (TP + TN)/Total = (90 + 63)/250 = 0.612 or, **61.20%**

Lift Score: 1.181. Lift Score > 1 indicates that the model can be useful.

ROC Curve

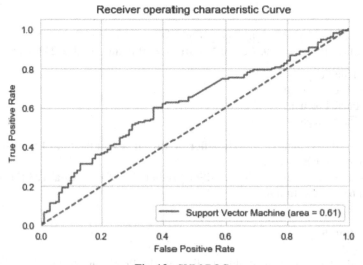

Fig. 13. SVM ROC

In the diagram above (Fig. 13), it can be seen that the **ROC curve is quite close to the baseline**. Thus, it can be inferred that this is not a good model but better than Model 1 (Logistics Regressions) for this classification.

Model 3: Decision Tree Classifier

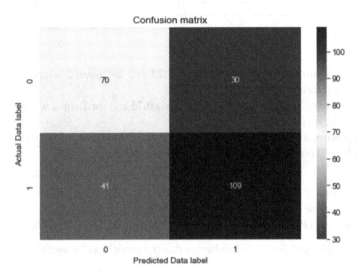

Fig. 14. Decision Tree Classifier Confusion Matrix

From the <u>Confusion Matrix</u> (Fig. 14), it can be seen that:
Actual Positives = 150, Actual Negatives = 100 and Total = 250. Out of which:

- TP = True Positives = 109
- TN = True Negatives = 70
- FP = False Positives = 30
- FN = False Negatives = 41 i.e.

- Out of all the predictions, the **classifier predicted yes** for the total **139 times**, out of which **109** were **actual yes**.
- Out of all the predictions, the **classifier predicted no** for the total **111 times**, out of which **70** were **actual no**.

Classification Report

The classification report (Table 5) displays the Precision, Recall, F1 and Support scores for the model.

- The **precision** for Team 1 winning the match is **0.78** and for Team 2 winning (or Team 1 losing) the match is **0.63**.

Table 5. Decision Tree Classifier Classification Report

	precision	recall	f1-score	support
0	0.63	0.70	0.66	100
1	0.78	0.73	0.75	150
accuracy			0.72	250
macro avg	0.71	0.71	0.71	250
weighted avg	0.72	0.72	0.72	250
Accuracy of decision tree classifier on test set:0.7160				

- **Recall** for Team 1 winning the match is **0.73** and for Team 2 winning (or Team 1 losing) the match is **0.70**.
- The **F1-score** of Team 1 winning the match is **0.75** and for Team 2 winning (or Team 1 losing) the match is **0.66**.
- **Support scores:** In the IPL data-set the data tested for Team 1 winning the match is **150** and for Team 2 winning (or Team 1 losing) the match is **100**.

Accuracy Score

The accuracy score of the model is 0. 7160, which is reasonable.

Accuracy = (TP + TN)/Total = (109 + 70)/250 = 0.716 or, **71.60%**

Lift Score: 1.31. Lift Score > 1 indicates that the model can be useful.

ROC Curve

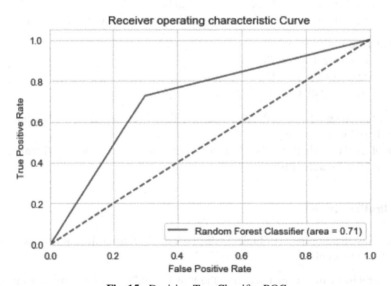

Fig. 15. Decision Tree Classifier ROC

In the diagram above (Fig. 15), it can be seen that the **ROC curve is away from the baseline towards the top-left portion of the graph**. Thus, it can be said that this is quite a reasonable model and better than Model 1 (Logistics Regression) and Model 2 (SVM) for this classification.

Model 4: Random Forest Classifier

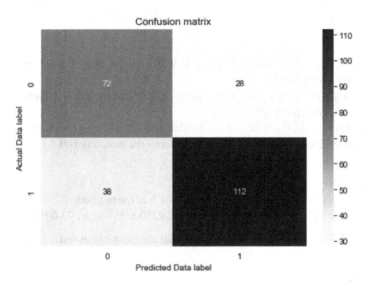

Fig. 16. Random Forest Classifier Confusion Matrix

From the above <u>Confusion Matrix</u> (Fig. 16), it can be seen that:
Actual Positives = 150, Actual Negatives = 100 and Total = 250 Out of which:

- TP = True Positives = 112
- TN = True Negatives = 72
- FP = False Positives = 28
- FN = False Negatives = 38 i.e.

- Out of all the predictions, the **classifier predicted yes** for the total **140 times**, out of which **112 were actual yes**.
- Out of all the predictions, the **classifier predicted no** for the total **110 times**, out of which **72 were actual no**.

Classification Report
The classification report (Table 6) displays the Precision, Recall, F1 and Support scores for the model.

- The **precision** for Team 1 winning the match is **0.80** and for Team 2 winning (or Team 1 losing) the match is **0.65**.

Table 6. Random Forest Classifier Classification Report

	precision	recall	f1-score	support
0	0.65	0.72	0.69	100
1	0.80	0.75	0.77	150
accuracy			0.74	250
macro avg	0.73	0.73	0.73	250
weighted avg	0.74	0.74	0.74	250
Accuracy of Random Forest classifier on test set: 0.7360				

- **Recall** for Team 1 winning the match is **0.75** and for Team 2 winning (or Team 1 losing) the match is **0.72**.
- The **F1-score** of Team 1 winning the match is **0.77** and for Team 2 winning (or Team 1 losing) the match is **0.69**.
- **Support scores**: In the IPL data-set the data tested for Team 1 winning the match is **150** and for Team 2 winning (or Team 1 losing) the match is **100**.

Accuracy Score
The accuracy score of the model is **0. 7360**, which is **quite good**.
    Accuracy = (TP + TN)/Total = (112 + 72)/250 = 0.736 or, **73.60%**

Lift Score: 1.333. Lift Score > 1 indicates that the model is useful.

ROC Curve

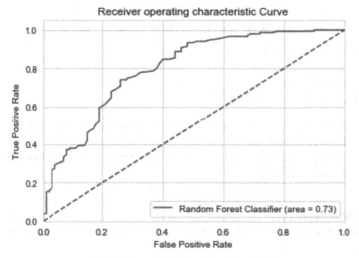

Fig. 17. Random Forest Classifier ROC

In the above diagram (Fig. 17), it can be seen that **the ROC curve is away from the baseline and is approaching the top-left section**. Thus, it can be said that this is quite a good model for this classification.

Comparing the four classification models, it can be seen that the **accuracy** of **Random Forest Classifier** (Model 4) is the **highest**. The other metrics used like Confusion Matrix, Lift Score and ROC Curve also show that the prediction ability **of Random Forest Classifier** is **better** than Logistic Regression, SVM and Decision Tree. Therefore, Random Forest Classifier is the model chosen for IPL Match winner prediction.

7 Validation

For cross-validation of the Random Forest model, a set of 8 random records were provided as input.

Input Data
See Tables 7 and 8.

Table 7. Input Data for Model Validation

Season	venue	team1	team2
14	MA Chidambaram Stadium, Chepauk	Kolkata Knight Riders	Sunrisers Hyderabad
14	Wankhede Stadium	Punjab Kings	Rajasthan Royals
14	MA Chidambaram Stadium, Chepauk	Mumbai Indians	Kolkata Knight Riders
14	MA Chidambaram Stadium, Chepauk	Royal Challengers Bangalore	Sunrisers Hyderabad
14	Wankhede Stadium	Delhi Capitals	Rajasthan Royals
14	Wankhede Stadium	Punjab Kings	Chennai Super Kings
14	Wankhede Stadium	Punjab Kings	Delhi Capitals
14	MA Chidambaram Stadium, Chepauk	Punjab Kings	Sunrisers Hyderabad

Table 8. Input Data for Model Validation (contd.)

toss_winner	toss_decision	result	result_margin	eliminator	method
Sunrisers Hyderabad	field	runs	10	N	NA
Rajasthan Royals	field	runs	4	N	NA
Kolkata Knight Riders	field	runs	10	N	NA
Sunrisers Hyderabad	field	runs	6	N	NA
Rajasthan Royals	field	wickets	3	N	NA
Chennai Super Kings	field	wickets	6	N	NA
Delhi Capitals	field	wickets	6	N	NA
Punjab Kings	bat	wickets	9	N	NA

Output: (team1_win is the Target Variable)
See Table 9.

From the Confusion Matrix (Fig. 18), it can be seen that:

Actual Positives = 4, Actual Negatives = 4 and Total = 8 Out of which:

Table 9. Model Output

Expected winner	Expected team1_win	Model Output
Kolkata Knight Riders	1	0
Punjab Kings	1	1
Mumbai Indians	1	0
Royal Challengers Bangalore	1	0
Rajasthan Royals	0	0
Chennai Super Kings	0	0
Delhi Capitals	0	0
Sunrisers Hyderabad	0	0

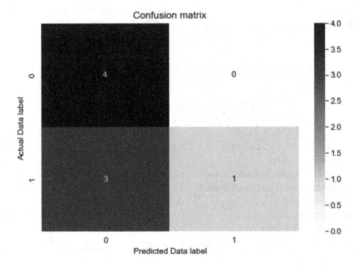

Fig. 18. Model Output Confusion Matrix (Random Forest Classifier)

- TP = True Positives = 1
- TN = True Negatives = 4
- FP = False Positives = 0
- FN = False Negatives = 3 i.e.

- Out of all the predictions, the **classifier predicted yes** for the total **1 time**, out of which **1 was actual yes**.
- Out of all the predictions, the **classifier predicted no** for the total **7 times**, out of which **4 were actual no**.

The input records had **4 cases of Team1 winning** the match and **4 cases of Team1 losing** the match. It can be seen that **out the 8 records, 5 records were correctly classified**. All the 4 records of Team1 losing the match were correctly identified whereas only 1 record of Team1 wining was identified correctly. 3 records of Team1 wining were wrongly classified as Team1 losing.

8   Conclusions

On the basis of IPL match history data from 2008 to 2021, classification model was built. From this dataset, the model was build using **features** like Season (1–14), venue, Team1, Team2, Toss_winner, Toss_decision (bat/field), Result (runs/wickets), Result_margin, Eliminator (Y/N), Method (NA or D/L) to predict the **target variable** team1_win (0/1).

Here four classification models – Logistic Regression, Support Vector Machine, Decision Tree and Random Forest were built in **Python** and checked for accuracy of the results. Out of these models, Random Forest Classifier had the best accuracy, lift score and ROC Curve. The **accuracy** of the Random Forest Classifier was found to be **73.60%** and the **Lift score** was **1.33** (>1). The **ROC Curve also extends away from the baseline towards the top-left section of the graph** indicating that it performs reasonably better than any purely random classifier. Also, this model can predict the outcome of a match with a better accuracy as compared to Singh and Kaur (2017) as studied in Literature Review section.

9   Future Scope of Research

This project can be improvised for future improvements and contribute towards building further enhanced models.

- This model can be used as tool to predict which team will win or lose a match, thus enabling teams to modify their strategies, generating money minting option for betting companies etc.
- This model can be extended to consider ball by ball analysis of every match. It will require team specific analysis to identify the conditions in which it is performing well and where it needs improvement. It will help a team analyse its player performance, overall team performance and formulate winning strategies.
- This model can be enhanced to include the data of individual players and their forms should also be analysed. Players of a team collectively compose a team. Therefore, analysing them at an individual level will help a team perform better.
- It can also be used to build better teams in fantasy leagues like Dream11.
- Other models can be developed and fine-tuned with additional features to improve the prediction of a match.

From here, it can be concluded that although this model considers each team on the whole and not on the basis of its player composition yet it is quite **reasonable** (as compared to other models) and can be used **to predict which of the two teams playing an IPL match will win**.

10   Limitations

This model has certain limitations.

- The features used to build the models consider a match as a whole i.e., for every season of IPL, the IPL teams and the conditions of the overall matches are considered.

- As mentioned earlier, IPL teams are contractual i.e., the players have a contract for a certain period with a specific team. Team compositions can change once the contract is over and fresh auctions take place. As the players of a team change, the overall team performance is also impacted.
- In some seasons, new IPL teams have been introduced. For example, in 2022, two new teams played. Historical data is unavailable for these new teams.
- Per team, ball by ball analysis was not done here. But such analysis can help to identify the strengths and weakness of the teams, the detailed strategies required to win a match etc. This in-depth analysis can help in understanding individual team performance and the performance of its players. This is beyond the scope of this project.
- Other factors like weather, form of a player, crowd support might also help in predicting a match outcome which was not considered here.

References

Kapadia, K., Abdel-Jaber, H., Thabtah, F., Hadi, W.: Sport analytics for cricket game results using machine learning: an experimental study. Appl. Comput. Inform. (2020)

Patel, N., Pandya, M.: IPL player's performance prediction. Int. J. Comput. Sci. Eng. **7**, 478–481 (2019)

Pathak, N., Wadhwa, H.: Applications of modern classification techniques to predict the outcome of ODI cricket. Procedia Comput. Sci. **87**, 55–60 (2016)

Prakash, C.D., Patvardhan, C., Lakshmi, C.V.: Data analytics based deep mayo predictor for IPL-9. Int. J. Comput. Appl. **152**(6), 6–10 (2016)

Sinha, A.: Application of Machine Learning in Cricket and Predictive Analytics of IPL 2020. Preprints 2020, 2020100436

Singh, S., Kaur, P.: IPL visualization and prediction using HBase. Procedia Comput. Sci. **122**, 910–915 (2017)

Vistro, D.M., Rasheed, F., David, L.G.: The cricket winner prediction with application of machine learning and data analytics. Int. J. Sci. Technol. Res. **8**(09) (2019)

Web References:

a. https://www.saedsayad.com/decision_tree.htm. Accessed 16 June 2021
b. https://www.javatpoint.com/machine-learning-support-vector-machine-algorithm. Accessed 16 June 2021
c. https://www.geeksforgeeks.org/understanding-logistic-regression/. Accessed 16 June 2021
d. https://www.kaggle.com/patrickb1912/indian-premier-league-2021-dataset. Accessed 16 June 2021
e. https://www.britannica.com/sports/cricket-sport. Accessed 21 June 2021
f. https://medium.com/   @saikatbhattacharya/model-evaluation-techniques-for-classification-models-eac30092c38b. Accessed 21 June 2021
g. https://machinelearningmastery.com/machine-learning-algorithms-mini-course/. Accessed 21 June 2021
h. https://machinelearningmastery.com/types-of-classification-in-machine-learning/. Accessed 21 June 2021
i. https://builtin.com/data-science/random-forest-algorithm. Accessed 20 June 2021

j. https://blog.betway.com/cricket/ipl-explained-everything-you-need-to-know/. Accessed 21
    June 2021
k. https://www.kaggle.com/sathyannarayan/predicting-outcome-of-ipl-match-based-on-variab
    les/data. Accessed 21 June 2021
l. https://www.datacamp.com/community/tutorials/random-forests-classifier-python. Accessed
    20 June 2021
m. https://www.analyticsvidhya.com/blog/2017/09/understaing-support-vector-machine-exa
    mple-code/. Accessed 16 June 2021
n. https://www.codespeedy.com/ipl-winner-prediction-using-machine-learning-in-python/.
    Accessed 17 June 2021

Knowledge Graph-Based Evaluation Metric for Conversational AI Systems: A Step Towards Quantifying Semantic Textual Similarity

Rajat Gaur[✉] and Ankit Dwivedi

Ernst and Young LLP, Technology Consulting Plot No. 67, Sector 44, Gurugram 122003, Haryana, India
rajatgaur625@gmail.com

Abstract. Machines face difficulty in comprehending the significance of textual information, unlike humans who can easily understand it. The process of semantic analysis aids machines in deciphering the intended meaning of textual information and extracting pertinent data. This, in turn, not only provides valuable insights but also minimizes the need for manual labor. Semantic Textual Similarity in Natural Language processing is one of the most challenging tasks that the research community faces. There have been many traditional methods to evaluate the similarity of two sentences that fall into categories of word-based metrics or embedding based metrics. In natural language understanding (NLU), semantic similarity refers to the degree of likeness or similarity in meaning between two or more pieces of text, words, phrases, or concepts. Evaluating semantic similarity enables the extraction of valuable information, thereby contributing significant insights while minimizing the need for manual labor. The ability of machines to understand the context just like humans do has been a challenging problem for so long. The objective of this research is to introduce a novel evaluation metric for measuring the textual similarity between two texts. The proposed metric will give us NERC scores based on a knowledge-graph approach that can be applied to assess the similarity between texts. The proposed evaluation metrics considers four different features that quantifies the similarity on the level of Nodes, Entities, Relationship and Context. The aggregate score considering the appropriate weights is calculated while considering each of these features in our proposed metrics to generate a final score ranging with least value 0 and maximum value 1. The evaluation was done on the Microsoft MSR Paraphrase Corpus dataset. Along with calculating a NERC score, other scores using already available metrics have also been calculated and reported and the results found during the experiment are comparable to the existing metrics.

Keywords: Knowledge Graph · Semantic Similarity · Natural Language Understanding · Wordnet · Textual Analytics · Named Entity Recognition

1 Introduction

In the era of Industry 4.0, with advancement in Conversational AI systems & Graph Databases [1], industries are leveraging Chatbots & other forms of conversational AI systems like Alexa, Siri, Google Assistant to engage with customers & giving multi-lingual 24 × 7 customer support. According to the MarketsandMarkets research, the global market for conversational AI is expected to exhibit a compound annual growth rate (CAGR) of 21.8% during the forecast period. The market is anticipated to reach an estimated value of USD 18.4 billion by 2026, up from an estimated USD 6.8 billion in 2021.

These Conversational AI systems directly interact with end uses & customers. Therefore, for a successful business leveraging these conversational systems, it is important to evaluate the accuracy of these systems in a semantic way and understand the meaning of conversation as humans do. As we know, Text Comprehension might be very intuitive for humans while for machines this task has always been very challenging. In Natural Language Generation tasks where the output is a human-like text, the correctness of the generated text and how similar it is to the actual human understanding, is often the objective.

The goal of any NLP evaluation metric is to determine how closely the Machine generated output matches a professional human produced text. If we take an example of an Image Caption Generator, then two different humans could describe the same image in two different manners without considering the same choice of words. However, the context would remain the same.

This is the basis of our proposed evaluation method. Via this study, we have proposed an automatic evaluation method that is used to find textual similarity between texts while considering the context as well. This method will help the researchers to come to conclusions more accurately. Furthermore, it will have a high correlation with human evaluation, which makes it more reliable in determining the performance of the evaluation metric.

The industry has seen increased usage of text analytics over a period of time and there have been many use cases that harness the power of Knowledge Graph. A Knowledge Graph is a concept that is used to represent knowledge in structured form. It consists of entities, which are things or concepts, and relationships, which describe how those entities are related to each other. Knowledge graphs have become an important tool in the field of NLP, as they enable the extraction and organization of information from unstructured data, such as text documents, into a structured form that can be easily queried and analyzed. By building a knowledge graph of entities and relationships related to a particular domain, such as movies, a search engine can provide more accurate and relevant results to user queries. For example, a search for "films starring Tom Hanks" can be easily answered by a knowledge graph that includes information about movies, actors, and their relationships. Another important application of knowledge graphs in industry is in the field of natural language understanding (NLU). NLU systems use knowledge graphs to represent the meaning of natural language sentences in a structured form. This enables the system to understand the intent behind the sentence and provide a relevant response. For example, a virtual assistant that uses a knowledge graph to represent information about restaurants can understand a sentence like "I want to find a good Italian restaurant in the city center" and provide a list of relevant restaurants. Knowledge graphs are also used in industry for knowledge management and data integration. By building a knowledge graph of entities and relationships related to a particular domain, organizations can better understand the relationships between different pieces of information and identify new insights and opportunities. For example, a pharmaceutical company can use a knowledge graph to represent information about drugs, diseases, and their relationships, and use this to identify new drug targets or potential side effects. Hence, knowledge graphs have become an important tool in the industry for NLP tasks, including search engines, NLU systems, and knowledge management. By representing knowledge in a structured form, knowledge graphs enable more accurate and relevant results to user queries and can help organizations identify new insights and opportunities. With our study, we aim to propose a metric that can help the industries to evaluate the similarity of texts based on top of knowledge graphs. This proposition could be even more helpful for those industries which have massive data in their Knowledge Graph Databases that will enable them to identify the similarity between their graphs by utilizing the semantic similarity approach proposed in this paper. Since, they have already converted the heavy text in a Knowledge Graph form, a need of quantification of semantic similarity in this context arises which suggests that instead of going with the traditional metrics to compare and evaluate similarity, this proposed approach can be easily employed on top of the already built Knowledge Graph databases to evaluate the performance metric of semantic textual similarity. However, in the application of the proposed metric in industry specific use cases, it is advisable to build up the dictionary for customization in reference to the use case so that the entities and relations are meaningfully captured and depicted by the Knowledge graph representation, on top of which the proposed metric can be employed to evaluate results. For instance, if the use case is in reference to Turkish language, a word net can be developed [2] that will further assist in the representation in the Knowledge Graph format and then NERC scores can be calculated to give an understanding about the semantic similarity. NERC scores refer to the similarity measures obtained by quantifying four distinct features on viz. Node level, Entity level, Relationship level, and Context level. This information is consistently applied throughout the paper.

2 Literature Review

In literature, there are various ways in which one can evaluate the similarity between two texts. Broadly speaking, the text similarity approach can be divided into four sections viz Corpus based, String based, Knowledge based and Hybrid based [3] and out of these 4 approaches, knowledge-based approach takes in account for semantic similarity and semantic relatedness between two texts. Figure 1 provides an overview of various methodologies employed in the research domain to evaluate the similarity between two texts.

Evaluating the quality of machine translation (MT) is a complex task that involves assessing multiple factors such as the accuracy, consistency, and fluency of the translation. There are various techniques and methodologies available for conducting human evaluations, as outlined by Reeder (2001). A few of them include techniques such as BLEU [4].

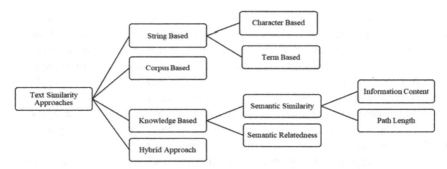

Fig. 1. Different approaches of calculating text similarity

2.1 Wordnet

Wordnet is basically a dictionary which is used to tackle disambiguation in text. It is a lexical database with tons of different words and sentences [5]. It is a large lexical database of English Words. The information between entities and relationships between those entities is correctly mapped in wordnet. It also stores various relations to a word which include synonyms, hypernym, holonym, meronym, etc. and contains forms of noun, verb, adjective, etc. Giving an example - "I ride a bike tomorrow", here intent is clear but grammatically incorrect hence we want to propose a metric based on the key information which is being captured.

WordNet includes several different similarity measures. For example, Wu & Palmer's similarity calculates similarity by considering the depths of the two synsets in the network, as well as where their most specific ancestor node (Least Common Subsumer (LCS)) is.

Nouns, verbs, adjectives, and adverbs are grouped into sets of cognitive synonyms called 'synsets', each expressing a distinct concept. Synsets are interlinked using conceptual-semantic and lexical relations such as hyponymy and antonymy.

Knowledge-Graph is an intelligent representation of raw text in the form of relationships and nodes. The earlier works that have been done in evaluation metrics harness the power of knowledge graph by relying on Word Sense Disambiguation (WSD).

Word Sense Disambiguation (WSD) is a task in computational linguistics that dates back to the 1940s, when machine translation was first developed [6]. The goal of WSD is to determine the appropriate meaning of a word given its surrounding context. This is a challenging task for machines, as humans are able to perform this task unconsciously. WSD is a classification task, where the system's goal is to assign a specific word to one of its senses as defined by a lexical dictionary. For example, the word "bank" has 18 different senses defined in the WordNet lexical database [7]. WSD systems can be divided into four main categories: supervised, semi-supervised, unsupervised, and knowledge-based. Supervised systems rely on a large sense-annotated training dataset. Semi-supervised systems leverage a bootstrapping technique by utilizing a small seed set of annotated training data in conjunction with a larger corpus of unannotated senses. On the other hand, unsupervised methods employ techniques such as context clustering, word clustering, and graph-based algorithms such as the PageRank algorithm. Knowledge-based approaches rely on a Knowledge Graph's structure and features, including taxonomic and non-taxonomic relations, a concept's Information Content, and paths. Of the four categories, supervised and knowledge-based approaches are the most promising.

WSD (Word Sense Disambiguation) is a computational task that aims to accurately classify a word into its appropriate sense, based on the context in which it is used. Since the 1940s, computational linguistics has devoted significant research efforts to develop various algorithms and techniques to address this task. However, due to discrepancies in sense choices across dictionaries and the complexity involved in compiling and annotating test datasets by humans, WSD poses a significant challenge. To overcome this difficulty, researchers often rely on extensive machine-readable lexical dictionaries such as WordNet, Wikipedia, and BabelNet.

2.2 Overlap Systems

The Lesk algorithm, which derives its name from its creator, is primarily centered on the similarity of words between a given context and the definition of a sense from a knowledge base. The sense with the most matching words is designated as the correct sense. However, the Lesk algorithm possesses certain limitations such as its sensitivity to an exact word match and the briefness of the definitions within WordNet. To overcome these limitations, Nanerjee and Padersen [8] expanded on the Lesk algorithm by incorporating associated concepts from the knowledge base. These associated concepts are identified based on direct relations with the candidate sense, such as hypernyms or meronyms.

2.3 Semantic Analysis

Semantic text similarity essentially means to compute similarity based on meaning. In NLU it aims to capture the meaning of the text while considering its context and grammar roles.

There are two types of Semantic Analysis:

1. Lexical - understanding meaning of each word
2. Compositional: understanding meaning of each word along with context. We understand how the combination of individual words form the meaning of the text.

Tasks involved in semantic analysis are as follows:

- Word Sense Disambiguation - the occurrence of words in different contexts carry different meanings, e.g., Apple can be referred to as a fruit in one sentence or an organization in another one.
- Relationship Extraction - NER (identifying entities and extracting relationships)

2.4 Semantic Similarity Approach

Since WordNet's inception, numerous metrics for assessing semantic similarity have been developed. These metrics operate on the premise that words that co-occur within a sentence should share a conceptual association in knowledge bases like WordNet.

Pedersen et al. [9] proposed a variant of the Lesk overlap approach by considering all feasible sentence combinations that can be formed by candidate senses within a contextual window consisting of words surrounding a target word. The Pedersen algorithm is founded on a semantic similarity score, and it can employ any semantic similarity measure. However, its outcomes are comparatively inferior to those of more contemporary approaches. For instance, a recent investigation by Mittal and Jain utilized an average of three semantic similarity measures [10], including the Wu and Palmer measure, the Leacock and Chodorow path-based metric, and a node counting distance metric. The average of these three metrics was adopted to determine the similarity between each sense of a polysemous word and its neighboring words.

2.5 Knowledge Based Approach

Other methods have used the structure of the knowledge base to create a sub-graph in order to find out the appropriate sense within a sentence. Navigli and Lapata [11] developed a method that involves creating a graph containing all possible combinations of senses for ambiguous words, with each node representing a sense of a word sequence and edges representing relationships between senses. The graph is then used to evaluate each node based on the shortest path measure to find the most suitable sense for each word in the context.

The concept of Knowledge-Based Similarity involves assessing the level of similarity between words by utilizing information extracted from semantic networks. Among the semantic networks, WordNet is the most widely used, which is a vast lexical database consisting of English words that are tagged as nouns, verbs, adjectives, and adverbs, and they are classified into sets of synonyms known as synsets, each representing a unique concept. To determine the degree of similarity between words, various similarity measures have been proposed, such as Resnik similarity, Jiang-Conrath similarity, and Lin similarity [12].

3   Methodology

In this section, a novel evaluation metric has been proposed, referred to as NERC, for quantifying the similarity between two sentences. The proposed metric incorporates different features present in a text to compute a final similarity score. The NERC metric consists of two main stages. In the first stage, Named Entity Recognition (NER) is performed on both sentences, along with identifying the relationships between the recognized entities. In the second stage, a Knowledge Graph (KG) is constructed for each sentence and the similarity between the two KGs is evaluated by computing the NERC score.

Four features are considered for quantification in the proposed metric, namely, nodes, entities, relationships, and context. The quantification of these features is done through specific formulas, which are discussed in the following sections. This metric is a comprehensive approach to evaluate the similarity of two sentences by considering the various features present in the text and quantifying them to obtain a final similarity score.

3.1 Node Level Similarity

In the node level similarity, we have quantified the number of nodes belonging to each of the knowledge graphs. This intuitively quantifies the length of the sentence/docs of the sentences. The formula for the node level similarity has been mentioned below:

$$Sim_N = \begin{cases} 1 & if \ |N_2 - N_1| = 0; \\ \frac{1}{|N_2 - N_1|} & if \ |N_2 - N_1| \neq 0; \end{cases} \tag{1}$$

where, N1 is the number of nodes of knowledge graph of sentence 1 and N2 is the number of nodes in knowledge graph of sentence 2.

3.2 Entity Level Similarity

In the entity level similarity, we are quantifying the presence of the key words as our entities in the knowledge graph. Intuitively, it means that similar entities will generate a higher score than dissimilar entities. The formula for the entity level similarity has been mentioned below:

$$Sim_E = \frac{2 \times depth\,(LCA)}{depth(entity_1) + depth(entity_2)} \tag{2}$$

3.3 Relationship Level Similarity

In relationship level similarity, we find the edges of the knowledge graphs of two given sentences and then measure the similarity between the two using Wu-Palmer Similarity [13]. This intuitively quantifies whether the edges carry a similar relationship in the two knowledge graphs or not. The formula for the calculation of the relationship level similarity score has been mentioned below:

$$Sim_R = \frac{2 \times depth\,(LCA)}{depth(relation_1) + depth(relation_2)} \tag{3}$$

where, depth (LCA) refers to the depth of the least common ancestor (LCA) of the two relations (relation 1 and relation 2) while depth(relation) refers to the depth of the synset associated with the relation in WordNet hierarchy.

3.4 Context Level Similarity

The context level similarity quantities the key idea discussed in the main text. The intuition behind considering the contextual similarity is that if the key idea in the given two sentences is same then it should give a higher similarity score. The formula for calculating the context level similarity has been mentioned below:

$$Sim_C = \frac{2 \times depth\,(LCA)}{depth(context_1) + depth(context_2)} \tag{4}$$

where, depth (LCA) refers to the depth of the least common ancestor (LCA) of the two contexts (context 1 and context 2) while depth(context) refers to the depth of the synset associated with the context in WordNet hierarchy.

3.5 NERC Similarity Score

After calculating each of the above scores viz. SimN, SimE, SimR and SimC, we calculate the NERC score by considering the weighted averages of each of these. The formula for calculating the final NERC score has been mentioned below:

$$Score_{NERC} = w1(Sim_N) + w2(Sim_E) + w3(Sim_R) + w4(Sim_C) \tag{5}$$

where, $w1$, $w2$, $w3$ and $w4$ are the weights given to Sim_N, Sim_E, Sim_R and Sim_C respectively.

4 Benchmark Dataset used in Research

The dataset that has been considered for our analysis is the MSR-Paraphrase dataset [14]. The MSR-Paraphrase dataset contains a total of 5,801 sentence pairs that have been manually annotated as either being semantically equivalent or not. The dataset was created by researchers at Microsoft Research and is often used as a benchmark for evaluating the performance of natural language processing models, particularly those that are focused on paraphrase identification and text similarity tasks.

The dataset is divided into two parts: a training set and a testing set. The training set contains 4,076 sentence pairs, while the testing set contains 1,725 sentence pairs. The sentence pairs in the training set have been annotated as either being semantically equivalent or not, while the sentence pairs in the testing set have also been annotated as being semantically equivalent or not.

The dataset is balanced in terms of the number of positive and negative examples, with 2,908 sentence pairs annotated as semantically equivalent and 1,168 sentence pairs annotated as not semantically equivalent in the training set. In the test set, 1147 sentence pairs are annotated as semantically equivalent while 578 sentence pairs annotated as semantically not equivalent.

The MSR-Paraphrase dataset has been widely used in the research community as a benchmark dataset for evaluating the performance of natural language processing models, particularly those that are focused on paraphrase identification and text similarity tasks. The dataset is available for download and use by researchers.

5 Experiments

In our analysis and proposition, both the training and testing datasets from the MSR Microsoft Paraphrase Corpus have been consolidated for further examination. The final dataset utilized for our analysis comprises a total of 5,801 sentence pairs. Among these, 3,900 sentence pairs are semantically equivalent, while 1,901 sentence pairs are semantically dissimilar.

In the experimental setup, the following similarity measures have been calculated which have been used in the evaluation of the performance of the proposed NERC metric.

1. **NERC -** The NERC metric is the proposition of this research work that generates a score between 0 to 1 and quantifies the textual similarity between two given sentences. This score has been taken as an aggregate of the four different features that we considered viz. Nodes, Entities, Relationships and Context. The formula used for calculating this score has already been discussed in the Methodology section. White calculating the NERC score, for each of the weights 0.25 weightage has been considered and equal importance has been given to each of the quantified features.

2. **BLEU1-** BLEU (Bilingual Evaluation Understudy) is a commonly used method for evaluating the quality of machine translation [4]. It compares the machine-generated translation to a set of reference translations, and assigns a score based on the degree of similarity between the two. The original BLEU score, BLEU-1, is calculated based on the number of unigrams (single words) that match between the machine translation and the reference translation. BLEU-1 is considered to be a simple and fast method for evaluating machine translation, but it has some limitations. For example, it does not consider the ordering of words in the translation, and it does not account for the context in which the words are used. Additionally, BLEU-1 is more sensitive to the length of the translation than other BLEU variants, making it less suitable for comparing translations of different length.

3. **BLEU2 -** Like the BLEU-1 metric which compares the machine-generated translation to a set of reference translations, BLEU-2 considers the similarity of bigrams (pairs of words) instead of just unigrams (single words). It is considered to be a more advanced method than BLEU-1, providing a more accurate evaluation of the coherence and fluency of the translation. It is commonly used in research and development of machine translation and often used alongside BLEU-1 for a more comprehensive evaluation.

4. **BLEU3-** Also known as BLEU-n = 3, is a variant of the BLEU evaluation method for machine translation. BLEU-3, like BLEU-1 and BLEU-2, compares the machine-generated translation to a set of reference translations, but it considers the similarity of trigrams (groups of three words) in addition to bigrams and unigrams. However, it still has some limitations, such as not accounting for word ordering and context beyond trigrams. It is often used alongside BLEU-1 and BLEU-2 to provide a more comprehensive evaluation of machine translation quality.

5. **BLEU4-** Also known as BLEU-n = 4, is a variant of the BLEU evaluation method for machine translation. BLEU-4, like BLEU-1, BLEU-2, and BLEU-3, compares the machine-generated translation to a set of reference translations, but it considers the similarity of four grams (groups of four words) in addition to trigrams, bigrams and unigrams. The idea behind BLEU-4 is that it can provide an even more accurate evaluation of machine translations by considering the context of the words at a deeper level. The four grams approach helps to evaluate the coherence, fluency, grammatical correctness, and lexical choice of the translation, which are important factors in determining the overall quality of the machine generated translation.

6. **Cosine Similarity-** Cosine similarity is a measure of similarity between two non-zero vectors of an inner product space [15]. It is a commonly used similarity measure in natural language processing and information retrieval. The cosine similarity is calculated by taking the dot product of two vectors and dividing it by the product of the magnitudes of the vectors.

Example: Textual Similarity scores for semantically equivalent sentences.

Sentence 1 - Yucaipa owned Dominick's before selling the chain to Safeway in 1998 for $2.5 billion.

Sentence 2 - Yucaipa bought Dominick's in 1995 for $693 million and sold it to Safeway for $1.8 billion in 1998.

Sentence 3 - Amrozi accused his brother, whom he called "the witness", of deliberately distorting his evidence.

Sentence 4 - Referring to him as only "the witness", Amrozi accused his brother of deliberately distorting his evidence.

Here, sentence 1 and sentence 2 are semantically not equivalent sentences i.e., Quality-0 sentences while sentence 3 and sentence 4 are semantically equivalent sentences i.e., Quality-1. The Table1 shows the scores for each of the two categories.

Table 1. Similarity scores for semantically equivalent and dissimilar sentences

Metrics	Similarity Score (Quality-0)	Similarity Score (Quality-1)
NERC	0.277	0.664
BLUE-4	0.534	0.767
BLEU-3	0.604	0.789
BLEU-2	0.684	0.815
BLEU-1	0.777	0.859
Cosine similarity	0.741	0.901

Similarly, Table 2 summarizes these scores that have been calculated for each of the 5801 sentence pairs. It summarizes the average values of each of the scores by the quality of the sentence. In Table 2, we have reported the mean scores, median scores and variance in scores for each of the two categories i.e., for semantically equivalent as well as semantically not equivalent sentences.

Table 2. Mean, Median and Variance of scores in semantically equivalent and dissimilar sentences

	Mean Scores		Median Scores		Variance in Scores	
Metrics	Quality 0	Quality 1	Quality 0	Quality 1	Quality 0	Quality 1
NERC	0.556	0.602	0.541	0.603	0.017	0.019
BLUE-4	0.547	0.676	0.545	0.681	0.010	0.015
BLEU-3	0.587	0.707	0.586	0.712	0.009	0.013
BLEU-2	0.639	0.744	0.639	0.749	0.008	0.010
BLEU-1	0.719	0.797	0.724	0.808	0.008	0.008
Cosine similarity	0.760	0.842	0.762	0.855	0.010	0.007

6 Results

For the whole corpus of 5801 sentences, we calculated NERC scores along with Bleu-1, Blue-2, Blue-3, Bleu-4 and cosine similarity. Intuitively how would a human evaluate two similar or dissimilar sentences. A human would be able to assess whether two sentences are referring to the same context or not, irrespective of the choice of words. Hence, it would make sense that for semantically similar sentences a higher score should be assigned since the context is the same while for two sentences which are not about the same context, a lower score should have been assigned due to different contexts.

On that basis, Table 3 summarizes the 5-point summary with respect to each of the scores discussed in Table 2, the 5-point summary viz, Min, 1st Quartile, Median, 3rd Quartile and Max.

Table 3. Point Summary of the Scores

	Bleu1	Bleu2	Bleu3	Bleu4	NERC	Cosine Similarity
Min	0.342	0.303	0.272	0.239	0.278	0.408
Quartile 1	0.703	0.629	0.577	0.535	0.47	0.745
Median	0.779	0.709	0.664	0.63	0.584	0.827
Quartile 3	0.847	0.79	0.758	0.733	0.685	0.894
Max	1	0.994	0.988	0.98	1	1

The above table suggests that the proposed metrics have been varying in the same fashion as the other existing metrics. This is also significantly evident in Fig. 2

The main goal of analyzing the 5-point summary for each metric was to determine if the proposed metric was providing meaningful results or if the results were random. This was done by comparing the scores of semantically equivalent sentence pairs with those that were not semantically equivalent. Additionally, it was important to ensure that the proposed NERC scores followed the same trend as existing metrics. Our findings indicate that the NERC scores are not random and that the inclusion of intuitive human-like features in the quantification process has contributed to generating meaningful scores that can be further studied. This is demonstrated in the Fig. 2 which shows that the trend of the scores is consistent across the different metrics.

The findings of the study indicate that the NERC scores effectively reflect the meaning of text when evaluating similarity scores. It should be noted, however, that our evaluation was conducted on a general dataset, and there may be instances where open source Named Entity Extraction models struggle to identify entities and relationships between sentences. Nonetheless, in industrial contexts where Knowledge Graphs are already in use for analytics, and entities and relationships are pre-established, the results of our proposed metric are expected to be more favorable. It is to be noted that the NERC metric weights have been uniformly set to 0.25, with equal weightage given to each of the four quantification measures: node, entity, relationship, and context. This represents a novel approach, and we suggest that the weights used in conjunction with this metric

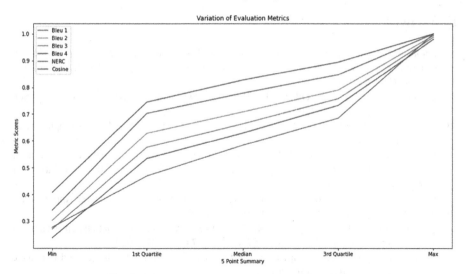

Fig. 2. Figure showing the trend of each of the scores

be reported alongside the NERC scores to ensure comparability of results. Therefore, we recommend reporting the weights used in NERC calculations for uniformity. While the weights have been set to 0.25 each for consistency, we acknowledge that the use case may warrant assigning more weight to certain features. If comparison across studies is desired, it is important to use consistent weights.

7 Conclusion

The objective of the research was to develop an intuitive evaluation measure for comparing texts, which takes into account various features that are considered by humans when understanding and comparing texts. These features include node level features such as sentence length, entity level features such as keyword similarity, relationship level features such as relation similarity and context level features such as the main idea or context. These scores were then combined using a weighted sum to generate the final NERC score, which indicates the semantic similarity between two sentences, with a score of 1 indicating the highest similarity and a score of 0 indicating that the sentences are not semantically equivalent. The first step in calculating the NERC scores is to construct a knowledge graph for each of the sentences being compared and then calculate the NERC score.

For future work, it would be beneficial to compare the proposed metrics against a human correlation dataset to determine its performance. While human correlation datasets are available at the document level, more research is needed to create a dataset of human correlation coefficients at the sentence level, which is necessary for our proposed metrics. Additionally, it would be valuable to extend the sentence level similarity measure to a document level measure to see if the proposed metrics can be applied to both sentence and document level, and if there are any limitations based on document length.

References

1. kumar Kaliyar, R.: Graph databases: a survey. In: International Conference on Computing, Communication & Automation, Greater Noida, India, pp. 785–790 (2015). https://doi.org/10.1109/CCAA.2015.7148480
2. Bilgin, O., Çetinoğlu, Ö., Oflazer, K.: Building a wordnet for Turkish. Rom. J. Inf. Sci. Technol. 7(1–2), 163–172 (2004)
3. Gomaa, W.H., Fahmy, A.A.: A survey of text similarity approaches. Int. J. Comput. Appl. 68(13), 13–18 (2013)
4. Papineni, K., Roukos, S., Ward, T., Zhu, W.J.: Bleu: a method for automatic evaluation of machine translation. In: Proceedings of the 40th Annual Meeting of the Association for Computational Linguistics, pp. 311–318 (2002)
5. Fellbaum, C.: WordNet. In: Theory and Applications of Ontology: Computer Applications, pp. 231–243. Dordrecht: Springer Netherlands (2010)
6. AlMousa, M., Benlamri, R., Khoury, R.: A novel word sense disambiguation approach using WordNet knowledge graph. Comput. Speech Lang. 74, 101337 (2022)
7. Jurafsky, D., Martin, J.H.: Chapter 19: word senses and wordnet. In: Speech and Language Processing. Third Edition draft (2018)
8. Banerjee, S., Pedersen, T.: An adapted Lesk algorithm for word sense disambiguation using WordNet. In: Gelbukh, A. (ed.) CICLing 2002. LNCS, vol. 2276, pp. 136–145. Springer, Heidelberg (2002). https://doi.org/10.1007/3-540-45715-1_11
9. Pedersen, T., Banerjee, S., Patwardhan, S.: Maximizing semantic relatedness to perform word sense disambiguation. Technical Report. Research Report UMSI 2005/25, University of Minnesota Supercomputing Institute (2005)
10. Mittal, K., Jain, A.: Word sense disambiguation method using semantic similarity measures and Owa operator. ICTACT J. Soft Comput. 5(2) (2015)
11. Navigli, R., Lapata, M.: Graph connectivity measures for unsupervised word sense disambiguation. In: IJCAI, pp. 1683–1688 (2007)
12. Seco, N., Veale, T., Hayes, J.: An intrinsic information content metric for semantic similarity in WordNet. In: Ecai, vol. 16, p. 1089 (2004)
13. Guessoum, D., Miraoui, M., Tadj, C.: A modification of wu and palmer semantic similarity measure. In: The Tenth International Conference on Mobile Ubiquitous Computing, Systems, Services and Technologies, pp. 42–46 (2016)
14. Dolan, B., Brockett, C.: Automatically constructing a corpus of sentential paraphrases. In: Third International Workshop on Paraphrasing (IWP2005) (2005)
15. Rahutomo, F., Kitasuka, T., Aritsugi, M.: Semantic cosine similarity. In: The 7th international student conference on advanced science and technology ICAST, vol. 4, no. 1, p. 1 (2012)

Are We Nearing Singularity? A Study of Language Capabilities of ChatGPT

Suparna Dhar[1] ✉ and Indranil Bose[2]

[1] NSHM Knowledge Campus, Kolkata 700053, India
suparna.dhar@nshm.com
[2] IIM Ahmedabad, Vastrapur, Ahmedabad, India

Abstract. There has been a huge uproar about ChatGPT in recent times. There are widespread fears that generative models such as ChatGPT will affect employability and bring humanity closer towards singularity. In this paper, we assessed the capabilities and limitations of ChatGPT empirically. We subjected ChatGPT to a series of questions with different levels of lexical and syntactic language complexity and manually scored the responses. We measured the lexical and syntactic complexity of the questions using text analytics techniques. The analysis of the responses showed ChatGPT as a highly capable language model, with certain limitations. ChatGPT showed an increased likelihood of generating an inaccurate response with increase in syntactic language complexity. Though there have been sporadic efforts and reports about ChatGPT's limitations, this paper presents a pioneering structured approach towards assessing ChatGPT's language capabilities. This study takes a small, yet important step in research towards assessing language models. It opens avenues for future research towards strengthening the generator model and developing the discriminator dataset.

Keywords: ChatGPT · Flesch Readability Ease Score · Language Complexity · Language Model · Logistic Regression

1 Introduction

There has been a huge uproar about ChatGPT in recent times. It has led to comments such as "Will #ChatGPT make #coding tests for #engineers obsolete?" on Twitter. The current uproar is a part of a larger discussion on whether machines can simulate human behavior. Allan Turing contended that machines can simulate human behavior closely [1]. Some people believe that in near future "computers are going to trump people. That they will be smarter than we are. Not just better at doing sums than us and knowing what the best route is to Basildon. They already do that. But that they will be able to understand what we say, learn from experience, crack jokes, tell stories, flirt" [2]. Some people claim that some powerful technologies are threatening to make humans an endangered species [3]. The Singularity hypothesis projects that technological advancements in artificial 'superintelligent' agents will radically change human civilization by the middle of the 21st century [4]. The world has already witnessed the emergence of several 'superintelligent'

S. Dhar et al. (Eds.): AGC 2023, CCIS 2008, pp. 125–135, 2024.
https://doi.org/10.1007/978-3-031-50815-8_8

technologies, such as metaverse, digital twins, augmented reality, virtual reality, mixed reality, extended reality, and many more [5]. New technologies and concepts, tools and techniques are on the rise. ChatGPT is one of the latest additions to the list.

ChatGPT is a natural language processing (NLP) chatbot powered by OpenAI's GPT-3 language model. It is designed to generate meaningful conversations and respond to user input in a natural and human-like way. ChatGPT can be used for customer service, customer engagement, and entertainment purposes.

Fig. 1. ChatGPT's response about itself

This uproar about technological singularity is further strengthened by huge employee layoffs by tech companies [6]. The Information Technology (IT) industry is shaken by the VUCA (volatility, uncertainty, complexity and ambiguity) environment ushered by the technological advancements [7]. On the other hand, there have been reports of inaccuracies in responses produced by ChatGPT. The question is, how far does ChatGPT take us towards singularity?

Though ChatGPT is a recent model, it has generated good academic interest. The literature on ChatGPT is sparse, but growing rapidly. Scholars have enumerated impact of ChatGPT on public health, learning and teaching, higher education, and other topics [8–10]. There has been a debate about the usage of ChatGPT in research [11–13]. Some scholars promoted the use of ChatGPT as a coauthor for scientific research papers [14]. Some scholars have tried to compare the intelligence level of ChatGPT with human using neuropsychological methods [15]. There is a gap is empirical research to test the language capabilities of ChatGPT. There was no literature on ChatGPT from the Singularity hypothesis perspective.

ChatGPT is a language model, based on OpenAI, that can generate human like answers to questions [16]. Figure 1 presents ChatGPT's response when asked about itself. An analysis of ChatGPT showed that model was able to handle questions with lower textual complexity well. The results were mixed when the model was exposed to texts with higher complexity. The model offered correct response when asked simple questions such as "Tell me about Kolkata" or "Tell me about the capital of West Bengal". However, the responses were not accurate when the model was asked questions with higher text complexity, such as, "Tell me about a city where Prince Dwarakanath Tagore's famous grandson was born". This led us to question ChatGPT's ability to answer questions having higher language complexity.

Our study aims to assess the language capabilities of ChatGPT. We assessed the capabilities and limitations of ChatGPT, by subjecting it to a series of questions with different levels of language complexity. The analysis of the responses showed an increased likelihood of ChatGPT generating an inaccurate response with increase in syntactic language complexity. The lexical complexity of the questions did not affect the accuracy of responses. Though there have been sporadic reports about ChatGPT's limitations, to the best of our knowledge, this is the first structured approach towards assessing ChatGPT's language capabilities. This paper presents the language modeling and language complexity concepts in Sect. 2, research methods in Sect. 3, the results in Sect. 4, discussions, contributions and future research directions, and limitations of the study in Sect. 5, and the conclusion in Sect. 6.

2 Language Complexity

In this section we present the concept of language complexity from the language modeling for AI perspective. Scholars proposed different techniques to measure language complexity. Flesch [17] proposed Flesch Reading Ease Score, a statistical measure of language complexity, which was one of the first objective measures of language complexity. The measure was used to grade language proficiency of students at different levels of education. When computers became popular, scholars focused on cognitive complexity of language. The objective was to minimize the cognitive complexity of language in human computer interaction in order to reduce computational complexity of the system [18]. Intelligent AI models are expected to solve cognitively complex tasks. The cognitive task complexity of language processing involves ability to deal with lexical complexity, syntactic complexity, and semantic complexity and generate responses [19, 20]. The measures have been borrowed from the field of linguistic studies, primarily used to measure language proficiency of students.

2.1 Lexical Complexity

Lexical complexity measures the complexity of words used in language. It involves measurement of rich lexicon, and word forms. Lexical sophistication refers to the use of a broad range of vocabulary [19]. Flesch readability ease score, given in formula (1) is a comprehensive measure of lexical complexity of natural language.

$$
\begin{aligned}
Flesch\,Reading\,Ease\,Score = {}& 206.835 \\
& - 1.015 * (Number\,of\,Words)/(Number\,Sentences) \quad (1) \\
& - 84.6 * (Total\,Syllables)/(Total\,Words)
\end{aligned}
$$

Other measures of lexical complexity include Flesch-Kincaid grade level score. Flesch-Kincaid grade level score maps the complexity level to language proficiency levels of school students in different grades.

2.2 Syntactic Complexity

Syntactic complexity deals with structural elements of text, such as structuration of the sentence, usage of phrases, formation of paragraphs, etc. It involves adherence to Grammar and writing styles. The length of the sentence, number of phrases per clause, length of the phrases, number of clauses in a sentence, and average length of clauses are measures of syntactic complexity [21]. Ability to manage syntactic complexity indicates language proficiency [22].

2.3 Semantic Complexity

The semantic layer of text is more tacit than the lexical and syntactic layers. Consequently, it is more difficult to measure and fewer scholars have attempted to model or measure sematic complexity of text data. Besharati & Izadi [20] proposed the intuitionistic Deciding About Semantic Complexity of a Text (DAST) model to measure semantic complexity. DAST posits that "semantic is considered as a lattice of intuitions and, as a result, semantic complexity is defined as the result of a calculation on this lattice" [20]. DAST offers a set of theoretic definitions for computing sematic complexity.

3 Methods

ChatGPT is built on OpenAI's GPT-3 family of language models [16]. The GPT-3 family of models include a) text-davinci-003, b) text-curie-001, c) text-babbage-001, and d) text-ada-001. All the GPT-3 models have been trained to understand and generate natural language, and have ability to handle different levels of language and task complexity. Text-davinci-003 is the latest and most capable of the GPT-3 family. We used the text-davinci-003 model to generate answers for 66 questions having different levels of language complexity. We assessed the correctness and the appropriateness of the answers manually. The accurate answers were score 1, and the inaccurate answers were scored 0. ChatGPT generated accurate answers for 44 questions, and inaccurate answers for 22 questions. When posed the question "Who is the monarch of England?", ChatGPT responded with "The current monarch of England is Queen Elizabeth II." The language models' responses were aligned to the information used to train the models. The models were not capable of maintaining contemporaneousness. For this analysis, we used time-invariant questions.

We used the language complexity measures Flesch reading ease score (readability), length of the question (number of words), number of clauses, average length of clauses, number of phrases, and subphrases as indicators of language complexity. We used an online calculator to measure the Flesch reading ease score (readability). We used a combination of NLTK parser, context free grammar, and manual scoring to score the syntactic language complexity measures.

We performed one-way ANOVA analysis to compare the mean values of the language complexity measures for the accurate and inaccurate responses [23]. In the next step, we tried to build a classification model to predict accurate response by ChatGPT. In our analysis, the output variable was binary (accuracy of response generated by ChatGPT coded as 0 for inaccurate response and 1 for accurate response) and the independent variables were either numeric (number of words, number of phrases, number of clauses, readability) or categorical (sub-phrases). We performed logistic regression analysis to find the expected change in log odds of having the outcome (accuracy of response) per unit change in the language complexity measures [24]. We used binary logistic regression in SPSS for the analysis. Logistic regression requires a minimum sample size of 10 per group of the dependent variable [25]. This justifies the use of Logistic Regression for the analysis with the collected sample. We split the data 80:20 with 20% holdout for validation.

4   Results

Table 1 presents sample results of our initial study of ChatGPT. Some more samples are provided in Appendix A.

Table 1. Answers to simple straightforward questions

Answers generated by text-davinci-003 model to questions posed by researchers	Assessment
Q1: Tell me about Kolkata A1: Kolkata, formerly known as Calcutta, is the capital of the Indian state of West Ben-gal. It is the second largest city in India and is located on the eastern bank of the Hooghly River. Kolkata is a major cultural, educational, and commercial center in India, and is known for its vibrant culture and colonial architecture. It is home to many universities, museums, and theaters, and is a major hub for the Bengali film industry. Kolkata is also known for its street food, which includes iconic dishes such as roshogolla, jhalmuri, and kathi rolls	Correct & appropriate
Q2. Tell me about the capital of West Bengal A2. The capital of West Bengal is Kolkata, formerly known as Calcutta. It is the largest city in the state and the seventh-most populous city in India. It is located on the east bank of the Hooghly River and is the commercial, cultural, and educational center of East India. Kolkata is known for its grand colonial architecture, art galleries, cultural festivals, and is considered the "cultural capital" of India. The city is home to numerous educational institutions, including the University of Calcutta, one of the oldest and most prestigious universities in India. Kolkata is also known for its delicious street food and is a popular tourist destination	Correct & appropriate
Q3. Tell me about a city where Tagore was born A3. Rabindranath Tagore was born in Calcutta, India, on May 7, 1861. Calcutta (now known as Kolkata) was at the time a bustling port city filled with rich cultural and literary activity. It was a center of the Bengali Renaissance and Tagore was surrounded by writers, intellectuals, and artists. The city was also home to a number of educational institutions, including the famous Hindu College, which Tagore attended. Tagore's upbringing in Calcutta would have a lasting impact on his life and work, and he would frequently return to the city throughout his life	Correct & appropriate
Q4. Tell me about a city where Prince Dwarakanath Tagore's famous grandson was born A4. Rabindranath Tagore, the famous Indian poet, artist, playwright, novelist, and composer, was born in Calcutta (now Kolkata), India on May 7, 1861. He was the grandson of Prince Dwarakanath Tagore, a wealthy and influential Bengali aristocrat. Tagore was born into an affluent and influential family in Calcutta, where he was exposed to a variety of cultural influences from an early age. He was educated at home by tutors, and was a prolific reader and writer from a young age. He wrote poetry and stories in Bengali, and later in English, and his works were widely acclaimed both in India and abroad. He was the first non-European to be awarded the Nobel Prize in Literature in 1913. Tagore was also an influential figure in India's cultural and political life, and his works continue to be widely read and admired today	Not accurate The answer talked about Tagore and not the city

One-way ANOVA analysis: The results of the ANOVA analysis are presented in Table 2. The mean values for number of words, number of clauses, number of phrases, and sub-phrases were higher for the inaccurate response class compared to the accurate response class. The mean difference between the classes was significant for number of words, number of clauses, number of phrases, and the sub-phrases only. Figure 2 depicts the distributions of (a) number of words, (b) number of clauses, and (c) number of phrases for the response classes. Higher readability (Flesch reading ease score) did not improve the accuracy of responses.

Table 2. Results of one-way ANOVA analysis

Measure	$Mean_{Accurate}$	$Mean_{Inaccurate}$	p-value
Number of words	10.61	15.55	<0.01
Number of clauses	1.57	2.14	<0.05
Average length of clauses	6.91	7.92	n.s
Number of phrases	0.93	1.86	<0.01
Sub-phrases	0.23	0.59	<0.01
Flesch reading ease score (readability)	74.65	76.38	n.s

Logistic regression: The data satisfied the assumptions for logistic regression. We performed logistic regression models with response class as the dependent variable, and number of words, number of clauses, number of phrases, and readability as independent variables. We eliminated the average length of clauses from this analysis to avoid multicollinearity. Overall model fit -2 Log Likelihood was 72.44. Hosmer and Lemeshow Test showed a significance level 0.871 indicating no significant difference between the predicted and actual value, so the model fit was acceptable. The Cox & Snell R Square was 0.228 and Nagelkerke R Square was 0.316. The model showed accuracy of 84.6 percent for the validation data. The (original) beta coefficient values were negative for all the independent variables indication that an increase in the value of the independent variable increased the likelihood of inaccurate response. The number of phrases ($\beta = -0.821$, p-value < 0.01) was significant.

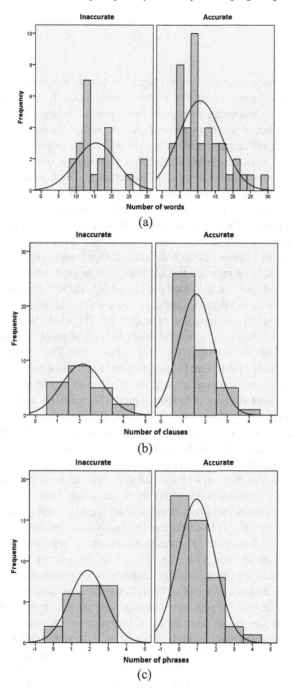

Fig. 2. Distribution of (a) number of words, (b) number of clauses, and (c) number of phrases for the response classes

5 Discussion

The mean comparison analysis showed higher syntactic language complexity for the inaccurate response classes compared to the accurate class. It indicates a likelihood of ChatGPT generating inaccurate responses with increase in language complexity of the question. The number of words, number of clauses, number of phrases, and sub-phrases showed significant difference between the means for the response classes. The mean lexical complexity measure readability (Flesch reading ease score) of the question and the average length of clauses did not show any change. The logistic regression showed significant negative association between the average length of clauses and the log odds of accurate answer – increase in this language complexity measure increased the likelihood of inaccurate response. The R Square values showed that the independent variables in the model explained more that 31 percent of variability in the response classes.

Contributions and future research directions: Our results supported that language complexity was likely to increase the likelihood of an inaccurate answer from Chat-GPT. The analysis showed an increased likelihood of ChatGPT generating an inaccurate response with increase in syntactic language complexity. The lexical complexity of the questions did not affect the accuracy of responses. The study inspires two important research directions. Research is needed towards improvement of ChatGPT language models to address the shortcomings reported in this paper. There is a need to develop a dataset with questions of different language complexity measures along with the complexity measures to test language models. Such a dataset would not only be useful for ChatGPT, but for improvement of language model for artificial intelligence in general. This is specifically important as models like ChatGPT generates unique responses every time, which makes application of ensemble models with in-built discriminators, such as general adversarial networks redundant for improvement of the generator model.

Limitations of the study: The study was performed with a small dataset hand-generated by the author. Since it was difficult to design the questions with high language complexity, a larger dataset was not within the scope of this study. As the size of the class with inaccurate responses was small, the class with accurate responses was not increased indiscriminately to maintain class balance. The Logistic Regression analysis was limited to identification of independent variables that impacted group (accurate and inaccurate) membership. This study used a handful of language complexity measures as independent variables. There are other language complexity measures that may be studied by future scholars. In absence of a tool or operationalizable method to compute, we could not include any measure for semantic language complexity in this study. Availability of tools to measure semantic complexity of language will further open the scope for research in assessing language models. With larger a dataset, scholars may try to build predictive models using machine learning techniques for binary classification such as Logistic Regression, Decision Tree, Random Forest, Support Vector Machine, and even deep learning techniques such as Artificial Neural Networks predict class membership.

6 Conclusion

The study made an effort to assess the accuracy of ChatGPT language model, reported to be one of the biggest disruptors of the day[1]. Despite using a small hand-generated dataset (using text-davinci-003 model in the GPT-3 library), the study identified one language complexity measure (average length of clauses) that affected the accuracy of the response. The distribution of number of phrases too showed significant difference for the response classes. The lexical complexity measure (readability) did not show any significant difference. This study takes a small, yet important step in research towards assessing language models. Though there have been sporadic reports about ChatGPT's limitations, this paper is a pioneer in presenting a structured approach towards assessing ChatGPT's language capabilities.

Appendix a: Sample Responses by ChatGPT

Question	ChatGPT response	Correctness
I was facing East and turned 180°, which direction am I facing now?	You are now facing West	Yes
How much is 135°plus 45°?	180°	Yes
I was facing East and turned 135° plus 45°, which direction am I facing now?	You are now facing South-Southeast	No
The afternoon sun is behind me - which direction am I facing?	You are facing west	No
I was facing East and turned 180° twice, which direction am I facing now?	You are now facing West	No
Name a fruit that starts with b	Banana	Yes
Name a fruit that starts with b and ends with y	Blueberry	Yes
Name a fruit that starts with b and ends with y and contains a in between	Banana	No
What is the king's son called?	The king's son is typically referred to as the crown prince or the prince	Yes
What is the king's son's mother called?	The king's son's mother is typically referred to as the queen or the queen mother	Yes
What is the king's son's mother's son called?	The king's son's mother's son is called the king's grandson	No

---

[1] https://www.thestreet.com/technology/chatgpt-is-the-new-disruptor-in-chief-but-theres-A-catch.

References

1. Turing, A.M.: Intelligent machinery, a heretical theory*. Philos. Math. **4**, 256–260 (1996). https://doi.org/10.1093/philmat/4.3.256
2. Cadwalladr, C.: Are the robots about to rise? Google's new director of engineering thinks so…, (2014). https://www.theguardian.com/technology/2014/feb/22/robots-google-ray-kur zweil-terminator-singularity-artificial-intelligence
3. Joy, B.: Why the future doesn't need us (2000). http://www.wired.com/wired/archive/8.04/joy.%0Ahtml
4. Eden, A.H., Steinhart, E., Pearce, D., Moor, J.H.: Singularity hypotheses: a scientific and philosophical assessment. Front. Collect. 1–13 (2012). https://doi.org/10.1007/978-3-642-32560-1
5. Lee, L.-H., et al.: All one needs to know about Metaverse: a complete survey on technological singularity, virtual ecosystem, and research agenda. J. Latex Class Files. **14**, 1–66 (2021). https://doi.org/10.48550/arXiv.2110.05352
6. Trueman, C.: Tech layoffs in 2023: A timeline (2023). https://www.computerworld.com/art icle/3685936/tech-layoffs-in-2023-a-timeline.html
7. Hanine, S., Dinar, B.: The challenges of human capital management in the VUCA era. J. Hum. Resour. Sustainab. Stud. **10**, 503–514 (2022). https://doi.org/10.4236/jhrss.2022.103030
8. Biswas, S.S.: Role of chat GPT in public health. Ann. Biomed. Eng. (2023). https://doi.org/10.1007/s10439-023-03172-7
9. Rudolph, J., Tan, S., Tan, S.: ChatGPT: bullshit spewer or the end of traditional assessments in higher education? J. Appl. Learn. Teach. **6** (2023). https://doi.org/10.37074/jalt.2023.6.1.9
10. King, M.R.: ChatGPT: a conversation on artificial intelligence, chatbots, and plagiarism in higher education. Cell. Mol. Bioeng. **16**, 1–2 (2023). https://doi.org/10.1007/s12195-022-007 54-8
11. Lahat, A., Shachar, E., Avidan, B., Shatz, Z., Glicksberg, B.S., Klang, E.: Evaluating the use of large language model in identifying top research questions in gastroenterology. Sci. Rep. **13**, 4164 (2023). https://doi.org/10.1038/s41598-023-31412-2
12. Pourhoseingholi, M.A., Hatamnejad, M.R., Solhpour, A.: Does ChatGPT (or any other artificial intelligence language tools) deserve to be included in authorship list? Gastroenterol Hepatol Bed Bench (2023). https://doi.org/10.22037/ghfbb.v16i1.2747
13. Lund, B.D., Wang, T.: Chatting about ChatGPT: how may AI and GPT impact academia and libraries? Library Hi Tech News. preprint (2023). https://doi.org/10.1108/LHTN-01-2023-0009
14. Dwivedi, Y.K., et al.: "So what if ChatGPT wrote it?" Multidisciplinary perspectives on opportunities, challenges and implications of generative conversational AI for research, practice and policy. Int. J. Inf. Manage. **71**, 102642 (2023). https://doi.org/10.1016/j.ijinfomgt.2023.102642
15. Loconte, R., Orrù, G., Tribastone, M., Pietrini, P., Sartori, G.: Challenging ChatGPT's "intelligence" with human tools: A neuropsychological investigation on prefrontal functioning of a large language model. SSRN. https://doi.org/10.2139/ssrn.4377371
16. OpenAI: Models. https://platform.openai.com/docs/models/gpt-3 (2023)
17. Flesch, R.: A new readability yardstick. J. Appl. Psychol. **32**, 221–233 (1948). https://doi.org/10.1037/h0057532
18. Bovair, S., Kieras, D.E., Poison, P.G.: The acquisition and performance of text-editing skill: a cognitive complexity analysis. Hum. Comput. Interact. **5**, 1–48 (1990). https://doi.org/10.1207/s15327051hci0501_1
19. Johnson, M.D.: Cognitive task complexity and L2 written syntactic complexity, accuracy, lexical complexity, and fluency: a research synthesis and meta-analysis. J. Second. Lang. Writ. **37**, 13–38 (2017). https://doi.org/10.1016/j.jslw.2017.06.001

20. Besharati, M., Izadi, M.: DAST Model: Deciding About Semantic Complexity of a Text (2019)

21. Pallotti, G.: A simple view of linguistic complexity. Second. Lang. Res. **31**, 117–134 (2015). https://doi.org/10.1177/0267658314536435

22. Lourdes, O.: Syntactic complexity measures and their relationship to L2 proficiency: a research synthesis of college level L2 writing. Appl. Linguist. **24**, 492–518 (2003). https://doi.org/10.1093/applin/24.4.492

23. Park, H.M.: Comparing group means: t-tests and one-way ANOVA using Stata, SAS, R, and SPSS (2009). https://hdl.handle.net/2022/19735

24. Kleinbaum, D.G., Klein, M.: Logistic Regression. Springer, New York, NY (2010). https://doi.org/10.1007/978-1-4419-1742-3

25. Hair, J.F., Black, W.C., Babin, B.J., Anderson, R.E.: Multivariate data analysis. Pearson, Noida, India (2013)

Thresholding Techniques Comparsion on Grayscale Image Segmentation

Vinay Kumar Nassa[✉] [iD] and Ganesh S. Wedpathak

Department of Computer Science and Engineering, Rajarambapu Institute of Technology,
Rajaramnagar, India
vinaykumar.nassa@ritindia.edu.in

Abstract. Image segmentation is a prime domain of computer vision backed by a huge amount of research involving both image processing-based algorithm and learning-based techniques. Due to this there is an upsurge in different segmentation technique from the research community. Various image segmentation techniques have their strength and weakness and some specific application are more geared up to some segmentation techniques. The automation systems like object detection, robotics and intelligent video analytics, do a lot of segmentation technique and hence there is need to evaluate the performance of these techniques. The paper implements the different types of segmentation techniques. Threshold techniques including like histogram thresholding, mean thresholding, edge thresholding, variable thresholding and percentile (P%-tile) exists. Algorithms are applied using MATLAB coding on the considered images. White Pixel Ratio (WPR) parameter is used for analysis of various methods. Similar to the theoretical concept, practical approach shows that WPR is better for histogram thresholding as compared to other techniques.

Keywords: Computer Vision · Segmentation · Image Processing · Edge · Histogram pixel · P- tile and WPR

1 Introduction

Image segmentation is a sub-domain of computer vision and digital image processing which aims at grouping similar regions or segments of an image under their respective class labels. Since the entire process is digital, a representation of the analog image in the form of pixels is available, making the task of forming segments equivalent to that of grouping pixels.

Image segmentation is an extension of image classification where, in addition to classification, we perform localization. Image segmentation thus is a superset of image classification with the model pinpointing where a corresponding object is present by outlining the object's boundary techniques.

Image processing in general is to extract the primitive objects in a scene. A step in image processing where the knowledge obtained about objects is used to interpretate high level objects is termed as scene analysis. No unique segmentation technique

S. Dhar et al. (Eds.): AGC 2023, CCIS 2008, pp. 136–146, 2024.
https://doi.org/10.1007/978-3-031-50815-8_9

exists for image processing. Instead various methods and techniques are used for image segmentation. Output of segmentation varies with various parameters like method used, heuristic rules, parameter values and technique used.

One of the most important steps towards image analysis is segmentation which involves identifying homogeneous regions in the image having same properties/features. Image segmentation is an important step in the process of Digital Image Processing (DIP). Usually the image segmentation is considered as a intermediate process between image preprocessing and image recognition stage [1]. Image segmentation is a method in which a digital image is broken down into various subgroups called Image segments which helps in reducing the complexity of the image to make further processing or analysis of the image simpler. Segmentation in easy words is assigning labels to pixels. All picture elements or pixels belonging to the same category have a common label assigned to them. For example: a problem where the picture has to be provided as input for object detection. Rather than processing the whole image, the detector can be inputted with a region selected by a segmentation algorithm. This will prevent the detector from processing the whole image thereby reducing inference time (Fig. 1).

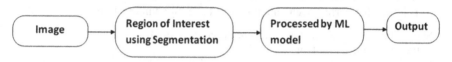

Fig. 1. Image processing model.

Segmentation is hard as well as difficult because:

- Homogeneity can be defined in terms of visual attributes as well as semantic boundaries.
- Due to diversity of attributes, problem formulation is difficult.

These are different techniques [1], edge detection [1], region based method [1], partial differential equation [2] and artificial neural network (ANN) [2]. Segmentation in general can be defined as partitioning the image pixels into classes depending upon pixel intensities as foreground and background.

In short by Image segmentation a given image boundary or region R is divided into a set of component regions Ri (Fig. 2).

Segmentation is the process of partitioning a digital image into multiple segments (sets of pixels, also known as super pixels [1]).

There are two broadly classification of segmentation techniques viz:

1. **Discontinuously based:** based on detecting edge points and linking them to form region boundaries. This approach relies on the discontinuity of pixel intensity values of the image. Line, Point, and Edge Detection techniques use this type of approach for obtaining intermediate segmentation results which can be later processed to obtain the final segmented image [3].
2. **Similarity based:** here the pixels which have similar visual attributes like intensity, color and texture are grouped together to form a segment. Machine Learning algorithms like clustering are based on this type of approach to segment an image.

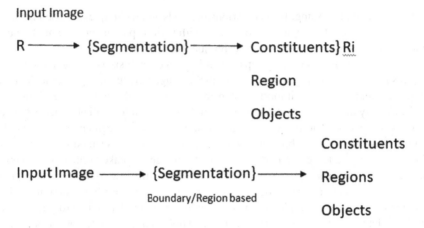

Fig. 2. Segmentation Method.

All such techniques involve formulating some homogeneity criteria.
It is found that correct segmentation must have following features:

- Should be able to extract the details required for output.
- Difficult and hard problem.
- Environment specific setting for image acquisition and processing need to be used. Techniques are application specific (Fig. 3).

$\left[\text{Discontinuity-based}\right]$ -------V/S----- $\left[\text{Similarity-based}\right]$

Edge/Boundary detection   Region Growing

Fig. 3. Segmentation Comparison Method.

2 Organisation of Paper

The paper is organized into various sections. Here the 3rd section discusses the basic concepts of Image thresholding and 4th section of threshold selection following 5th section as various thresholding techniques introduction. The results obtained by MAT-LAB CODE are explained in the 6th section following conclusion. Simplest way of segmentation is threshold base classification of image pixels in predefined classes is observed.

3 Image Thresholding

Image thresholding segmentation is a simple form of image segmentation. It is a way to create a binary or multi-color image based on setting a threshold value on the pixel intensity of the original image.

Simplest way of image thresholding is to segment an image into various regions based upon intensity and comparison of each pixel with a suitable threshold value. Pixel is assigned to one of the class labels based upon comparison. Depending upon threshold value chosen, threshold methods vary as Global techniques, Adaptive (Local techniques) and variable thresholding (Split, merge and grow) technique. Homogeneity and geometrical proximity are the basic features involved in all techniques.

Depending upon threshold value above techniques segments the pixels.

In Global thresholding threshold value is applied equally to all pixels in the image. Images having bimodal histogram having distinct valley separating modes are suitable for Global threshold based segmentation.

A disadvantage of this type of threshold is that it performs really poorly during poor illumination in the image.

To overcome the effect of illumination, the image is divided into various sub regions, and all these regions are segmented using the threshold value calculated for all these regions.

In Adaptive thresholding the threshold value is taken as intensity of neighbor pixel for the next pixel comparison.

Variable thresholding involves each pixel have its own different threshold value for comparison (Fig. 4).

$$T = T[x, y, p(x, y), f(x, y)] \tag{1}$$

where T is the threshold value $p(x, y), f(x, y)$ are points the gray level image pixels. Final labeling result g(x,y).(SINGLE THRESHOLD T)

$$g(x, y) = \begin{cases} 1 \, iff(x, y) > T \\ 0 iff(x, y) \le T \end{cases} \tag{2}$$

Input Image(R) ⟶ {Thresholding} ⟶ Labelled Image

Fig. 4. Image Thresholding

Final labeling result g (x,y) (TWO THRESHOLD T1, T2)

$$g(x, y) = \begin{cases} a iff(x, y) > T2 \\ bT1! iff(x, y) \le T2 \\ c iff(x, y) \le T2 \end{cases} \tag{3}$$

4 Threshold Selection

The most important in thresholding process is to find or decide the threshold value (or values). Several methods to choose a threshold value exist. Success of thresholding depends upon width and depth of valleys separating the histogram peaks.

Valleys in the image are dependent on several factors like variances and illumination gradient in the image. Large variances indicate that there is no pronounced valley in the histogram. Similarly absence of illumination gradient results in no pronounced valleys in histogram. There are various ways to select the threshold values. One such is user can manually choose a value or a thresholding algorithm can compute a value automatically, which is known as automatic thresholding [3, 6, 8].

Another way is to choose the mean and median values, which means that if objects pixels are brighter than the background, they should be also brighter than average.

This method is best suitable for wireless images with uniform background. However this will generally not the case always.

Another sophisticated approach might be to prepare a histogram of the image pixel intensities and use valley points as the threshold and employ the valley techniques. However this method employs large computational effort and expensive.

However in some cases image histogram may not have clearly defined valley points. In short, it is difficult often to select the accurate threshold value.

5 Various Thresholding Techniques

Based upon the selection of threshold values, various image thresholding techniques are classified as:

5.1 Percentile(P-tile)Thresholding

It is one of the earliest methods, which is based upon grey level histogram. It is generally termed as automatic threshold method which uses the knowledge about area size of the desired object. In this if the prior knowledge about the object and image is known then the first P% of area below the histogram is considered as threshold. This fixed percentage of picture area is known as P%. The threshold is defined as the grey level [5].

5.2 Mean Thresholding

As the name implies this method make use of mean value of the pixels as the threshold value. This method is employed strictly in cases where images have half of the pixels that belong to object and other half to the background. This technique case rarely happens [3, 6, 8].

The algorithm is explained as

1. Choose an initial value of threshold to classify the pixels to belong any of two groups depending upon that the pixel's intensity is greater or less than threshold.
2. Mean value is computed for each of the group and update the value which is average of mean values of intensity for the two groups.
3. Using the updated value of threshold again classify the pixels.
4. Repeat the process till the value of the threshold stabilizes between successive iterations.

Features of mean thresholding:

- The output of the method depends upon the initial threshold.
- Method is simple but time consuming.
- Means are to be computed at every iteration (Time complexity increases with size of the images)

5.3 Histogram Thresholding (HDT)

The disadvantage of mean thresholding (Iterative) is the increase in time complexity by increasing the number of iterations in large sized images.

This can be overcome by estimating the threshold value that separates the two homogenous region of the objects and background of an image. This is known as histogram technique.

This requires that, the image formation be of two homogenous separated regions and there exists a threshold value that separates' two segments [5]. The HDT is suitable when all regions of objects and background are homogenous except the area between objects and background.

We compute Pi as the fraction of pixels in the image which have the intensity level i

$$Pi = \frac{Ni}{MN} Pi > 0, 0 \le i \le L - 1 \tag{4}$$

for given L intensity levels, MN is the number of pixels. For a normalized histogram we have

$$\sum_{i=0}^{L-1} Pi = 1 \tag{5}$$

5.4 Edge Thresholding

Some times in the images for ground object of interest is so small that have negligible effect in the histogram due to domination by background intensities. Then use of histogram technique in this case, is difficult to separate both background and foreground regions. Since the Computed threshold will simplify divide the background region itself into two classes and it is edge to extract the edge pixels in the image by applying suitable edge conditions. Edge pixels generally belong to the boundary region of the foreground regions. Here similar to histogram techniques Computation is done only to the pixels near to edge pixels. Since in this technique boundary of a region or object that is closed is detected, here no. of objects of interest is equal to the number of boundaries in an image [7, 8].

5.5 Variable Thresholding

In variable thresholding separate threshold value is assigned for each pixel to classify (into one of the given classes) for different set of pixels [9, 10].

Here the images are divided into patches and each patch is given a separate threshold depending upon path contents. Generally the threshold value is taken to be weighted sum of mean intensity value of a path and global threshold value. Here better results are obtained in case illuminated and reflectance varies across image.

Pixel classification in variable thresholding can be adopted by defining some predicate function θ which may be a function of local parameters (and may be thresholds)

$$g(x,y) = \begin{cases} 1 \text{ if } \theta(\text{Local Parameter}) \text{ is True} \\ 0 \text{ if } \theta(\text{Local Parameter}) \text{ is False} \end{cases} \qquad (6)$$

6 Experimental Results and Discussion

Recently, there has been lots of interests in evaluating image segmentation techniques. Motivation for this is to optimize the existing techniques. Thus a no of segmentation techniques are developed to aid embedded/autonomous system to apply for a particular image [4]. Though no single approach can be said to be best, some methods perform better for some images then others. Segmentation techniques are broadly classified into subjective evaluation and objective evaluation depending upon whether human being has evaluate the image usually or not [11].

The whole image is divided into patches of same size. An image of 512 Ö512 is divided into eight equal patches. Binarisation of each patch is done by using any of thresholding method using MATLAB.

The ratio of the foreground image pixels to the background image pixels is known as White Pixel Ratio. This parameter is selected for comparison since each color image can be represented in binary form as Gray Scale Image. The method with high WPR is a good thresholding method.

6.1 Tank Image

(Fig. 5) (Table 1)

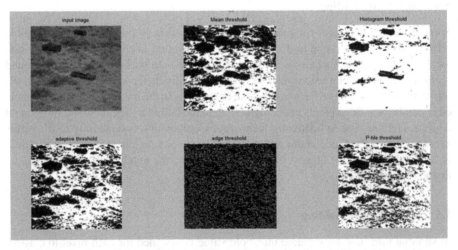

Fig. 5. Comparison of thresholding techniques of Tank image from left to right Input image.

Table 1. WPR of Tank Image.

Thresholding techniques	Black to white pixel ratio	WPR (%AGE)
Mean	113663:148481	56.64
Histogram	24660:237484	90.05
Edge	224499:37645	14.36
Variable	107002:155142	59.18
P-tile	80919:181225	69.13

6.2 Couple Image

(Fig. 6) (Table 2)

Fig. 6. Comparisons of thresholding techniques of Couple image from left to right Input image.

Table 2. WPR of Couple Image.

Thresholding techniques	Black to white pixel ratio	WPR (%AGE)
Mean	121031:141113	53.83
Histogram	43302:218842	83.48
Edge	235688:26456	10.09
Variable	127356:134788	51.41
P-tile	118882:143262	54.65

6.3   Girl Image

(Table 3)

Table 3. WPR of Girl Image.

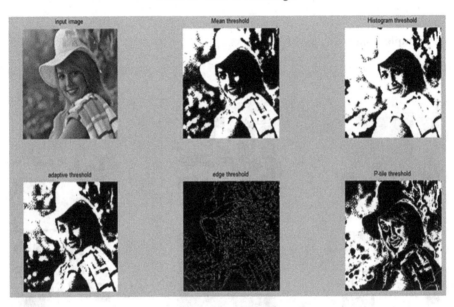

Thresholding techniques	Black to white pixel ratio	WPR (%AGE)
Mean	133411:128733	49.10
Histogram	7313:189031	72.10
Edge	243815:18329	6.99
Variable	131864:130280	49.69
P-tile	162299:99845	38.08

6.4   Plane Image

(Fig. 7) (Table 4).

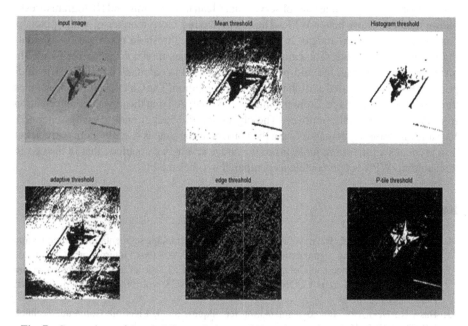

Fig. 7. Comparison of thresholding techniques of Plane image from left to right Input image

Table 4. WPR of Plane Image.

Thresholding techniques	Black to white pixel ratio	WPR (%AGE)
Mean	98994:163150	62.23
Histogram	9988:252156	96.18
Edge	238387:23757	9.06
Variable	87898:174246	66.46
P-tile	254746:7398	2.82

7   Applications of Image Segmentation

Image segmentation is an important step in artificial vision. Machines need to divide visual data into segments for segment-specific processing to take place.

Image segmentation thus finds its way in prominent fields like Robotics, Medical Imaging, Autonomous Vehicles, and Intelligent Video Analytics.

Apart from these applications, Image segmentation is also used by satellites on aerial imagery for segmenting out roads, buildings, and trees.

8 Conclusion

In this paper various segmentation techniques are evaluate in MATLAB based on the pixels. From the above tables it is observed that Mean thresholding and Histogram thresholding methods show better WPR. The image having more number of white pixels is considered good binary image. In mean thresholding, threshold is chosen by calculating mean of pixel value. The calculated mean becomes threshold value for image. In histogram thresholding, the mid-point method is used to calculate threshold. The mid-point method is also called as global thresholding method. The Calculated value becomes threshold for whole image. From the images it is analyzed that the unwanted information from the image is easily removed. The object and the background is easily separated. Variable thresholding manages to get the borders of the shapes slightly more correct, but also produces a little more junk. From the above results we conclude that in histogram thresholding WPR is appreciable as compared to other methods.

References

1. Sapna, S.V., Navin, R., Ravindar, P.: Comparative study of image segmentation techniques and object matching using segmentation. In: Author, F., Author, S (eds.) International Conference on Methods and Models in Computer Science (ICM2CS) Title of a proceedings paper. In: Editor, F., Editor, S. (eds.) CONFERENCE 2016, LNCS, vol. 9999, pp. 1–13. Springer, Heidelberg (2016)
2. Dass, R., Devi, P.S.: Image segmentation techniques. Int. J. Electron. Commun. Technol. (IJECT) 3(1) (2012)
3. Al-amri, S.S., Kalyankar, N.V., Khamitkar, S.D.: Image segmentation by using thresholding techniques. J. Comput. 2(5) (2010). ISSN 2151–9617
4. Goldmann, L., et al.: Towards fully automatic image segmentation evaluation. In: Blanc-Talon, J., Bourennane, S., Philips, W., Popescu, D., Scheunders, P. (eds.) Advanced Concepts for Intelligent Vision Systems. ACIVS 2008. Lecture Notes in Computer Science, vol. 5259, pp. 566–577 (2008). https://doi.org/10.1007/978-3-540-88458-3_51
5. Tobias, O.J., Sear, R.: Image segmentation by histogram thresholding using fuzzy sets. IEEE Trans. Image Process. 11(12) (2002)
6. Henden, P.C.: Exercise in Computer Vision A Comparison of Thresholding Methods –NTNU 20th November 200
7. Samopa, F., Asano, A.: Hybrid image thresholding method using edge detection. In: IJCSNS International Journal of Computer Science and Network Security, vol. 9, no. 4, pp. 292–2991 (2009)
8. Zuva, T., Olugbara, O.O., Ojo, S.O., Ngwira, S.M.: Image segmentation, available techniques, developments and open issues. Can. J. Image Process. Comput. Vis. 2(3) (2011)
9. Huang, Z.-K., Chau, K.-W.: A new image thresholding method based on gaussian mixture model. Appl. Math. Comput. 205(2), 899–907 (2008)
10. Banimelhem, O., Yahya, Y.A.: Multi-Thresholding Image Segmentation Using Genetic Algorithm. Segmentation: Region and Boundary Information Integration. Springer, pp. 408–422 (2011)
11. Zhang, H., Fritts, J.E., Goldman, S.A.: Image segmentation evaluation: a survey of unsupervised methods. Comput. Vis. Image Underst. 110(2008), 260–280 (2008)

Automated Pneumonia Diagnosis from Chest X-rays Using Deep Learning Techniques

Pratyay Ranjan Datta[✉] and Moumita Roy

NICA, Kolkata, West Bengal, India
pratyaydata@gmail.com

Abstract. Pneumonia is a common and potentially life-threatening respiratory infection. Accurate and timely diagnosis is critical for effective treatment, but traditional diagnostic methods can be time consuming and require specialized expertise. Deep learning, a subset of artificial intelligence, has shown promise in improving pneumonia diagnosis through analysis of medical imaging. The novelty of our research stems from the unique approach we used to collect our dataset of normal and pneumonia patients. Rather than relying on pre-existing datasets, we employed a novel data scraping method to collect data from Bing images. By utilizing this novel data collection method, we were able to overcome limitations of existing datasets, such as potential biases and limited sample sizes, and obtain a more representative and robust dataset for our study. Our dataset, thus, offers a unique and valuable resource for studying the relationship between pneumonia and various patient factors, providing new insights and opportunities for research in this field. This innovative data collection approach sets our study apart and contributes to the novelty and rigor of our research findings. In this study, we trained and evaluated five deep learning models (CNN, VGG16, ResNet50, DenseNet169 and InceptionV3) on a dataset of chest X-ray images for pneumonia diagnosis. We found that all models achieved high accuracy and demonstrated strong precision and recall values. The VGG16, ResNet50, and DenseNet169 models showed promising results with testing accuracies of 90%, 91%, and 90%, respectively. The InceptionV3 model had a lower testing accuracy of 89% . The precision-recall curve analysis showed high AUC values for all models, indicating good performance in detecting pneumonia. Our findings suggest that deep learning models can be effective tools for pneumonia diagnosis, with the potential to improve accuracy and efficiency of traditional diagnostic methods. However, further research is needed to address limitations and challenges, such as potential bias in dataset selection and limitations in generalizability. With careful consideration and validation, deep learning models have the potential to play a valuable role in improving pneumonia diagnosis and ultimately, patient outcomes.

Keywords: Pneumonia · Deep Learning · Chest X-ray images · Convolutional Neural Networks · VGG16 · ResNet50 · DenseNet169 · InceptionV3

© The Author(s), under exclusive license to Springer Nature Switzerland AG 2024
S. Dhar et al. (Eds.): AGC 2023, CCIS 2008, pp. 147–169, 2024.
https://doi.org/10.1007/978-3-031-50815-8_10

1 Introduction

1.1 Background and Motivation

Pneumonia is a respiratory infection that can lead to serious health complications and is a leading cause of death worldwide, especially in low- and middle-income countries. Early detection and accurate diagnosis of pneumonia is critical for effective treatment and management of the disease. However, traditional diagnosis methods, such as chest radiography and clinical examination, are often time-consuming, expensive, and require significant expertise, leading to delays in diagnosis and treatment. The emergence of computer vision and deep learning techniques has opened up new possibilities for automated and accurate pneumonia diagnosis. In recent years, various deep learning models have been proposed for pneumonia detection using chest X-ray images, achieving high accuracy and efficiency. In this paper, we propose a deep learning-based approach for pneumonia diagnosis using chest X-ray images. Our approach leverages the power of convolutional neural networks (CNNs) and transfer learning to automatically identify pneumonia patterns in chest X-rays. The proposed approach has the potential to significantly improve the accuracy and speed of pneumonia diagnosis, and thereby improve patient outcomes.

1.2 Overview of Pneumonia

Pneumonia is a respiratory infection caused by bacteria, viruses, or fungi. It affects the air sacs in the lungs, leading to inflammation and fluid accumulation, which can cause symptoms such as cough, fever, and difficulty breathing. Pneumonia can range in severity from mild to life threatening, and can have serious health consequences, particularly in vulnerable populations such as young children, the elderly, and people with weakened immune systems. The symptoms of pneumonia are widespread and can include pain, coughing, shortness of breath, and more. According to statistics, pneumonia affects about 7.7% of the world's population annually [1]. The categorization of pneumonia is primarily based on the underlying pathogenesis. Popular classification methods include infectious and non-infectious pneumonia. Infectious pneumonia, in turn, is classified into several subcategories such as bacterial, viral, mycoplasmas, and chlamydial pneumonia, among others. On the other hand, non-infectious pneumonia is further classified into immune-associated pneumonia, aspiration pneumonia caused by physical and chemical factors, and radiation pneumonia. These classification methods are widely used in the medical field to accurately diagnose and treat pneumonia [2]. In 2016, it was estimated that pneumonia caused approximately 16% of the 5.6 million deaths of children under the age of five, resulting in the deaths of approximately 880,000 children [3]. The fatality rate of pneumonia is closely linked to age, and the incidence of pneumonia surges significantly with age, particularly in individuals above the age of 65.

1.3 Importance of Early Diagnosis

Early detection and accurate diagnosis of pneumonia is critical for effective treatment and management of the disease. Delayed diagnosis and treatment can lead to serious complications and poor health outcomes. In addition, timely diagnosis can help to prevent the spread of infectious agents and reduce the burden on healthcare systems. However, traditional pneumonia diagnosis methods, such as chest radiography and clinical examination, are often time consuming, expensive, and require significant expertise, leading to delays in diagnosis and treatment.

1.4   Role of Computer Vision and Deep Learning

The emergence of computer vision and deep learning techniques has opened up new possibilities for automated and accurate pneumonia diagnosis. Deep learning models, particularly convolutional neural networks (CNNs), have shown great potential for automated analysis of medical images, including chest X-rays. These models can learn complex patterns and features in the images that are difficult for human experts to detect. By leveraging the power of deep learning and computer vision, automated pneumonia diagnosis can be achieved with high accuracy and efficiency, potentially improving patient outcomes and reducing the burden on healthcare systems.

2   Problem Statement

2.1   Limitations of Traditional Pneumonia Diagnosis Methods

Traditional pneumonia diagnosis methods, such as chest radiography and clinical examination, have several limitations that can lead to delayed diagnosis and treatment. These methods require significant expertise, are often time-consuming and expensive, and may not always provide accurate results. For instance, chest radiography can be difficult to interpret due to the overlap of pneumonia features with those of other respiratory diseases, leading to high false-negative or false positive rates.

2.2   Need for Automated and Accurate Diagnosis

There is a need for automated and accurate pneumonia diagnosis methods that can overcome the limitations of traditional methods. Computer vision and deep learning techniques offer a promising solution to this problem by enabling automated analysis of medical images, such as chest X-rays. Deep learning models, particularly CNNs, have shown great potential for accurate and efficient pneumonia diagnosis. By leveraging the power of deep learning and computer vision, pneumonia diagnosis can be achieved with high accuracy and efficiency, potentially improving patient outcomes and reducing the burden on healthcare systems.

3   Objectives

The main objective of this research is to develop a deep learning-based automated pneumonia diagnosis system using chest X-ray images. The system will aim to achieve high accuracy and efficiency in detecting and localizing pneumonia in the images.

4   Contributions

The proposed approach for pneumonia diagnosis using computer vision and deep learning is novel in several ways. Firstly, it leverages state-of-the-art CNN architectures, including VGG16, ResNet50, DenseNet169, and InceptionV3 to achieve high accuracy in detecting and localizing pneumonia in chest X-ray images. The proposed approach has the potential to impact clinical practice by providing an automated and accurate pneumonia diagnosis system that can improve patient outcomes and reduce healthcare costs. By leveraging the power of computer vision and deep learning, the proposed approach can achieve high accuracy and efficiency in pneumonia diagnosis, leading to earlier and more effective treatment, reduced hospitalization, and improved patient outcomes.

5   Literature Review

This section aims to provide a comprehensive overview of the current state of research in this area, including the strengths and limitations of existing studies, and identify any gaps or areas for future research. This literature review helps to establish the context and significance of the study, and enables the researcher to situate their own findings within the broader scholarly conversation on the topic. Lakhani et al. [4] developed a deep convolutional neural network (CNN) to classify pulmonary tuberculosis. In addition to their CNN, transfer learning models including AlexNet and GoogleNet were also utilized to classify chest X-ray images. To evaluate the performance of their models, the dataset was divided into training, testing, and validation sets, representing 68%, 14.9%, and 17.1% of the data, respectively. To improve model performance, the researchers employed data augmentation and pre-processing techniques, resulting in an impressive area under the curve (AUC) value of 0.99. Moreover, the model achieved a precision of 100% and a recall of 97.3%, further demonstrating its robustness and accuracy in tuberculosis classification. Sammy V. Militante et al. [5] used flexible and powerful deep learning methodologies to predict whether a patient has pneumonia by analyzing chest X-ray images. Six CNN models, including GoogLeNet, LeNet, VGG-16, AlexNet, StridedNet, and ResNet50, were employed to monitor the accuracy of each version learned. The models were trained using a dataset of 28,000 images with a decision size of 224 × 224 and batch sizes of 32 and 64, utilizing Adam as an optimizer with an adjusted 1e–4 learning rate and a 500-epoch. During the model development process, GoogLeNet and LeNet achieved a 98% accuracy rate, VGGNet-16 achieved a 97% accuracy rate, AlexNet and StridedNet models had an average accuracy rate of 96%, and the ResNet-50 model had an accuracy rate of 80%. The GoogleNet and LeNet models had the highest average accuracy for overall performance training. All six models were able to identify and predict pneumonia illness, including those with normal chest X-ray images. Ayan and Unver [6] applied VGG16 and Xception deep learning models to classify pneumonia using chest X-ray images. During the training phase, the models were fine-tuned after transfer learning. The performance of both networks was evaluated using various metrics. The VGG16 model achieved an accuracy of 87%, while the Xception model had an accuracy of 82%. The results showed that the Xception model performed better in detecting pneumonia cases, while the VGG16 model demonstrated good performance in identifying normal cases. Pan et al. [7] utilized an ensemble of Inception-ResNet v2, XceptionNet, and DenseNet-169 models to detect pneumonia and achieved the top score in the challenge with an mAP value of 0.33. However, we have not found any evidence that ensemble models have been used for classification tasks in the pneumonia detection problem. Therefore, we employed ensemble learning for the first time in this domain to classify lung X-rays into "Pneumonia" and "Normal" categories. To form the ensemble, we utilized three state-of-the-art CNN models with transfer learning, namely GoogLeNet, ResNet-18, and DenseNet-121, and a weighted average probability technique was used, where the weights were assigned using a novel approach. M Mujahid et al. [8] proposed an automatic pneumonia detection approach using chest X-ray images with deep learning-based ensemble models. The approach combines pre-trained models with a custom-built CNN model to achieve reduced training time and higher accuracy. The Inception-V3 ensemble model achieved the highest accuracy with a score of 99.29%, and 98.83%, 99.73%, and 99.28% scores for precision, recall, and F1, respectively. The ResNet50 ensemble obtained 98.93% accuracy, followed by the VGG-16 ensemble with a 98.06% accuracy score. The study shows that individual pre-trained models exhibit poor performance as compared to their ensemble models. In conclusion, several studies have utilized deep learning techniques to develop computer-aided systems for pneumonia detection using chest X-ray images. These studies have shown promising results, with some achieving high accuracy and outperforming radiologists in pneumonia detection. The use of deep learning models like CN, and data augmentation and transfer learning have all contributed to improving the performance of these systems. However, there is still room for improvement, and future research can focus on improving the robustness and generalizability of these models, as well as exploring the potential of other deep learning techniques in pneumonia detection.

6 Methodology

6.1 Data Scrapping

To demonstrate the novelty of our research on pneumonia detection using chest X-ray images, we used a unique approach to collect our dataset. Instead of using existing datasets from platforms like Kaggle, we employed data scraping techniques to extract image URLs from Bing's image search results. This allowed us to collect a diverse dataset with images of pneumonia and normal chest X-rays that hasn't been used before. By using this novel approach, we were able to achieve more accurate results in our machine learning models for pneumonia detection. We ensured that our data scraping methods were legal, ethical, and compliant with relevant regulations and guidelines, and took steps to protect the privacy and confidentiality of the collected data. Highlighting these aspects helps us demonstrate the novelty and value of our research, which may have implications for improved diagnosis and treatment of pneumonia using chest X-ray images.

6.2 Dataset Description

A total of 594 images were collected and used for training and testing the deep learning models. The dataset was divided into two main folders - training and testing, each containing two subfolders, one for normal cases and the other for pneumonia cases. The training set consisted of 418 images, with 209 normal and 209 pneumonia cases. The testing set consisted of 176 images, with 88 normal and 88 pneumonia cases (Fig. 1).

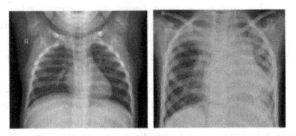

Fig. 1. Normal vs. Pneumonia

Preprocessing: Preprocessing Before being fed into the deep learning models, the chest X-ray images underwent several preprocessing steps, including:

- Resizing the images to a fixed dimension of 224 × 224 pixels to ensure that all the images had the same dimensions
- Normalizing the pixel values of the images to have a range of 0 to 1
- Creating a validation dataset by randomly selecting 20% of the images from the training dataset to monitor the models' performance during training and prevent overfitting. These preprocessing techniques were applied to ensure that the input data fed into the deep learning models were standardized and optimized for the models' performance.

6.3   Deep Learning Architecture and Models

Deep Learning Architecture and Models Used To detect pneumonia from chest X-ray images, several pre-trained deep learning models were used, including Convolutional Neural Networks (CNN), VGG16, ResNet50, DenseNet169, and InceptionV3.

CNN is a type of neural network that is commonly used for image classification tasks. VGG16, ResNet50, DenseNet169, and InceptionV3 are pre-trained CNN models that have been trained on large datasets such as ImageNet to extract meaningful features from images. It is a type of deep neural network that is widely used for image classification tasks which consists of multiple convolutional and pooling layers that can learn complex features from the input images. They are widely used in computer vision tasks, such as image classification, object detection, and segmentation.

CNNs consist of several layers, each of which performs a specific task. The first layer is typically a convolutional layer, which applies a set of filters (also called kernels or weights) to the input image to extract features such as edges, corners, and textures. Each filter slides over the input image, performing a dot product between its weights and the corresponding pixels in the input.

After the convolutional layer, the output is typically passed through a non-linear activation function, such as the Rectified Linear Unit (ReLU), which introduces non-linearity to the model. This is followed by a pooling layer, which reduces the spatial resolution of the output by taking the maximum or average value in each local region of the feature map.

The process of convolution and pooling is repeated several times, with each layer learning increasingly complex features. The final output is typically a fully connected layer, which maps the learned features to the output classes. The network is trained using a loss function, such as cross-entropy, to minimize the difference between the predicted output and the true labels. Here, is the summary of the model (Fig. 2).

```
Model: "sequential"
_____
 Layer (type)                Output Shape              Param #
=================================================================
 conv2d (Conv2D)             (None, 222, 222, 32)      896

 max_pooling2d (MaxPooling2D  (None, 111, 111, 32)     0
 )

 conv2d_1 (Conv2D)           (None, 109, 109, 64)      18496

 max_pooling2d_1 (MaxPooling  (None, 54, 54, 64)       0
 2D)

 conv2d_2 (Conv2D)           (None, 52, 52, 128)       73856

 max_pooling2d_2 (MaxPooling  (None, 26, 26, 128)      0
 2D)

 conv2d_3 (Conv2D)           (None, 24, 24, 128)       147584

 max_pooling2d_3 (MaxPooling  (None, 12, 12, 128)      0
 2D)

 flatten (Flatten)           (None, 18432)             0

 dense (Dense)               (None, 512)               9437696

 dense_1 (Dense)             (None, 2)                 1026

=================================================================
Total params: 9,679,554
Trainable params: 9,679,554
Non-trainable params: 0
_____
```

Fig. 2. CNN Model

VGG16 is a popular pre-trained deep learning model that has achieved state-of-the-art performance on several image classification benchmarks. The VGG16 network consists of 16 layers, which include 13 convolutional layers, 5 max pooling layers, and 3 fully connected layers. The convolutional layers are grouped into blocks, with each block containing multiple convolutional layers followed by a max pooling layer. The fully connected layers at the end of the network map the learned features to the output classes.

One of the key contributions of the VGG16 architecture is the use of very small convolutional filters (3×3) in all convolutional layers. This small filter size reduces the number of parameters in the network and enables the network to learn more local features. In addition, the use of max pooling layers after each convolutional block helps to reduce the spatial resolution of the features and introduce some degree of translation invariance to the network.

The VGG16 network is typically pre-trained on a large dataset such as ImageNet, which contains millions of images and thousands of categories. This pre-training helps the network to learn general features that can be transferred to other tasks, such as object detection and segmentation. Here, is the summary of the model (Fig. 3).

```
Model: "vgg16"
_____
 Layer (type)                Output Shape              Param #
=================================================================
 input_2 (InputLayer)        [(None, 224, 224, 3)]     0

 block1_conv1 (Conv2D)       (None, 224, 224, 64)      1792

 block1_conv2 (Conv2D)       (None, 224, 224, 64)      36928

 block1_pool (MaxPooling2D)  (None, 112, 112, 64)      0

 block2_conv1 (Conv2D)       (None, 112, 112, 128)     73856

 block2_conv2 (Conv2D)       (None, 112, 112, 128)     147584

 block2_pool (MaxPooling2D)  (None, 56, 56, 128)       0

 block3_conv1 (Conv2D)       (None, 56, 56, 256)       295168

 block3_conv2 (Conv2D)       (None, 56, 56, 256)       590080

 block3_conv3 (Conv2D)       (None, 56, 56, 256)       590080

 block3_pool (MaxPooling2D)  (None, 28, 28, 256)       0

 block4_conv1 (Conv2D)       (None, 28, 28, 512)       1180160

 block4_conv2 (Conv2D)       (None, 28, 28, 512)       2359808

 block4_conv3 (Conv2D)       (None, 28, 28, 512)       2359808

 block4_pool (MaxPooling2D)  (None, 14, 14, 512)       0

 block5_conv1 (Conv2D)       (None, 14, 14, 512)       2359808

 block5_conv2 (Conv2D)       (None, 14, 14, 512)       2359808

 block5_conv3 (Conv2D)       (None, 14, 14, 512)       2359808

 block5_pool (MaxPooling2D)  (None, 7, 7, 512)         0

=================================================================
Total params: 14,714,688
Trainable params: 14,714,688
Non-trainable params: 0
```

Fig. 3. VGG-16 Model

ResNet50 is another pre-trained deep learning model that is known for its ability to overcome the problem of vanishing gradients in very deep neural networks. The ResNet50 architecture consists of 50 layers, which includes 49 convolutional layers and 1 fully connected layer. The convolutional layers are organized into blocks, with each block containing multiple convolutional layers followed by a batch normalization layer and a ReLU activation function. Here, is the summary of the model (Fig. 4).

DenseNet169 is a deep learning model that has a unique architecture where each layer is connected to every other layer in a feed-forward fashion. This architecture promotes feature reuse and reduces the number of parameters in the model. The DenseNet169 architecture consists of 169 layers, which includes 166 convolutional layers and 3 fully connected layers. The convolutional layers are organized into blocks, with each block

```
Layer (type)                    Output Shape              Param #
===================================================================
resnet50 (Functional)           (None, 2048)              23587712

flatten_10 (Flatten)            (None, 2048)              0

batch_normalization (BatchN     (None, 2048)              8192
ormalization)

dense_20 (Dense)                (None, 2048)              4196352

batch_normalization_1 (Batc     (None, 2048)              8192
hNormalization)

dense_21 (Dense)                (None, 1024)              2098176

batch_normalization_2 (Batc     (None, 1024)              4096
hNormalization)

dense_22 (Dense)                (None, 2)                 2050

===================================================================
Total params: 29,904,770
Trainable params: 6,306,818
Non-trainable params: 23,597,952
```

Fig. 4. ResNet50 model

containing multiple layers that are densely connected to each other. The final output is passed through a global average pooling layer and a fully connected layer to produce the classification result.

DenseNet169 has achieved state-of-the-art performance on several image classification benchmarks, including ImageNet and CIFAR-10. The pre-trained DenseNet169 model is available in many deep learning frameworks and can be fine-tuned on new datasets for various computer vision tasks, such as object detection and segmentation. Here, is the summary of the model (Fig. 5).

Inception V3 is a deep learning model that uses multiple convolutional filters of different sizes to capture features at different scales. It has achieved state-of-the-art performance on several image classification benchmarks. One of the key features of Inception V3 is its use of "Inception modules", which are composed of a set of parallel convolutional filters of different sizes. By using filters of different sizes, the model is able to capture features at multiple scales, which helps to improve its ability to recognize objects of different sizes and shapes in images. In addition to its use of Inception modules, Inception V3 also incorporates other advanced techniques, such as batch normalization, which helps to speed up training, and the use of global average pooling, which helps to reduce the number of parameters in the model. Overall, Inception V3 has achieved state-of-the-art performance on several image classification benchmarks, including the ImageNet Large Scale Visual Recognition Challenge (ILSVRC), where it achieved a top-5 error rate of just 3.46%. This makes it one of the most accurate image classification models available today. Here, is the summary of the model (Fig. 6).

```
Layer (type)                 Output Shape         Param #
=================================================================
densenet169 (Functional)     (None, 1664)          12642880

flatten_12 (Flatten)         (None, 1664)          0

batch_normalization_6 (Batc  (None, 1664)          6656
hNormalization)

dense_26 (Dense)             (None, 2048)          3409920

dropout (Dropout)            (None, 2048)          0

batch_normalization_7 (Batc  (None, 2048)          8192
hNormalization)

dense_27 (Dense)             (None, 1024)          2098176

dropout_1 (Dropout)          (None, 1024)          0

batch_normalization_8 (Batc  (None, 1024)          4096
hNormalization)

dense_28 (Dense)             (None, 2)             2050
=================================================================
Total params: 18,171,970
Trainable params: 5,519,618
Non-trainable params: 12,652,352
```

Fig. 5. DenseNet169 model

By using a variety of deep learning models, this research paper explores the effectiveness of different architectures and models in diagnosing pneumonia from chest X-ray images. The results of this study can help to inform the development of more accurate and reliable diagnostic tools for pneumonia.

6.4 Hyper Parameter Tuning and Optimization

In this research paper, we performed hyperparameters tuning to identify the optimal combination of hyperparameters for each deep learning model. We varied the hyperparameters systematically, including the number of epochs, learning rate, and batch size.

After trying several combinations of hyperparameters, we found that the combination of 10 epochs, a learning rate of 0.0001, and a batch size of 15 worked best for our models. We selected these hyperparameters based on their ability to achieve high accuracy, precision, recall, and F1- score, as well as their ability to prevent overfitting.

The number of epochs determines the number of times the deep learning model iterates over the training data. By setting the number of epochs to 10, we ensured that the model had enough iterations to learn the patterns in the data without overfitting.

The learning rate determines the step size for updating the weights during training. By setting the learning rate to 0.0001, we ensured that the model learned gradually and did not overshoot the optimal weights.

```
Layer (type)                    Output Shape           Param #
=================================================================
inception_v3 (Functional)       (None, 5, 5, 2048)     21802784

global_average_pooling2d (G     (None, 2048)           0
lobalAveragePooling2D)

batch_normalization (BatchN     (None, 2048)           8192
ormalization)

dense (Dense)                   (None, 256)            524544

dropout (Dropout)               (None, 256)            0

batch_normalization_1 (Batc     (None, 256)            1024
hNormalization)

dense_1 (Dense)                 (None, 128)            32896

dropout_1 (Dropout)             (None, 128)            0

dense_2 (Dense)                 (None, 2)              258

=================================================================
Total params: 22,369,698
Trainable params: 562,306
Non-trainable params: 21,807,392
```

Fig. 6. Inception V3 model

The batch size determines the number of samples processed in each iteration. By setting the batch size to 15, we ensured that the model could learn from small batches of data efficiently while still being able to capture the important patterns in the data.

6.5 Performance Evaluation Metrics

In this research paper, we used several performance evaluation metrics to assess the effectiveness of our deep learning models in diagnosing pneumonia from chest X-ray images. The performance evaluation metrics we used included accuracy, precision, recall, and F1-score.

Accuracy: Accuracy measures how often the model is correct. It is defined as the number of correct predictions divided by the total number of predictions.

Accuracy= (TP + TN)/(TP + TN + FP + FN)

Precision: Precision measures how many of the positive predictions are actually positive. It is defined as the number of true positives divided by the sum of true positives and false positives.

Precision = TP/(TP + FP)

Recall: Recall measures how many of the actual positive cases are predicted as positive. It is defined as the number of true positives divided by the sum of true positives and false negatives.

Recall = TP/(TP + FN)

F1-score: F1-score is a harmonic mean of precision and recall, which provides a balanced measure of the two. It is defined as 2 times the product of precision and recall divided by the sum of precision and recall.

F1-score = 2 * ((Precision * Recall)/(Precision + Recall))

We also used confusion matrices to visualize the performance of our models. Confusion matrices show the number of true positive, true negative, false positive, and false negative predictions for each class. Overall, by using multiple performance evaluation metrics, we were able to assess the effectiveness of our models in diagnosing pneumonia accurately and reliably from chest X-ray images. The performance evaluation metrics and confusion matrices provided us with valuable insights into the strengths and weaknesses of our models, allowing us to improve our models further.

7   Results

7.1   Convolutional Neural Network (CNN)

We developed and trained a CNN model on a dataset of chest X-ray images to diagnose pneumonia. The model was trained for 10 epochs using the Adam optimizer, without a learning rate schedule. At the last epoch, the model achieved a training accuracy of 97.31% and a validation accuracy of 91.57%, with a loss of 0.0946 and a validation loss of 0.3570. The training and validation accuracy vs loss graph can be seen below (Fig. 7):

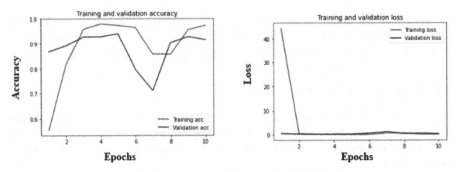

Fig. 7.   Accuracy vs Loss graph for CNN

We evaluated the model on a testing dataset, where it achieved an accuracy of 90%. The model correctly classified 77 cases of pneumonia as pneumonia and 82 cases of normal as normal. However, it also misclassified 11 cases of pneumonia as normal and 6 cases of normal as pneumonia. This classification report shows the performance of a model in detecting pneumonia and normal cases from a total of 176 cases. The precision and recall scores for pneumonia and normal cases are both high, indicating that the model is good at correctly identifying both classes. The accuracy of the model is also high at 0.90, meaning it correctly classified 90% of the cases. The classification report and confusion matrix can be seen below (Fig. 8 and (Fig. 9):

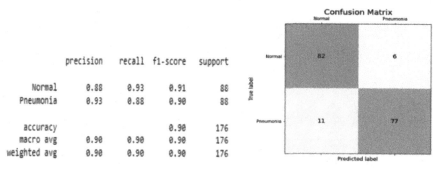

Fig. 8. Confusion matrix and Classification report for CNN

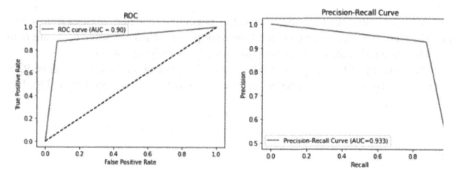

Fig. 9. ROC and Precision-recall curve

To further evaluate the performance of the model, The Receiver operating characteristic (ROC) curve for the model was calculated to be 0.92, indicating that the model performed well in distinguishing between positive and negative cases. Similarly, precision-recall curve and calculated the area under the curve (AUC), which was found to be 0.9333. The high AUC value indicates that the model has a good balance between precision and recall, making it an effective tool for diagnosing pneumonia from chest X-ray images.

7.2 CNN with Hyper parameter Tuning

We performed hyper parameter tuning on our CNN model by adding a dropout layer with a rate of 0.5 and adjusting the learning rate. The model was trained for 10 epochs using the Adam optimizer, without a learning rate schedule. At the last epoch, the model achieved a training accuracy of 92.24% and a validation accuracy of 95.18%, with a loss of 0.1838 and a validation loss of 0.1251.

The training and validation accuracy vs loss graph can be seen below (Fig. 10):

We evaluated the performance of the tuned model on the testing dataset, where it achieved an accuracy of 92%. The model correctly classified 80 cases of pneumonia as pneumonia and 82 cases of normal as normal. However, it misclassified 8 cases of pneumonia as normal and 6 cases of normal as pneumonia. This classification report

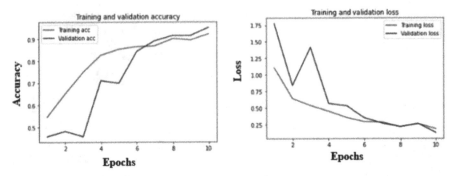

Fig. 10. Accuracy vs Loss graph for CNN with hyper parameter tuning

shows the performance of a model in detecting pneumonia and normal cases from a total of 176 cases. The precision and recall scores for pneumonia and normal cases are both high, indicating that the model is good at correctly identifying both classes.

The accuracy of the model is also high at 0.92, meaning it correctly classified 92% of the cases. The classification report and confusion matrix can be seen below (Fig. 11):

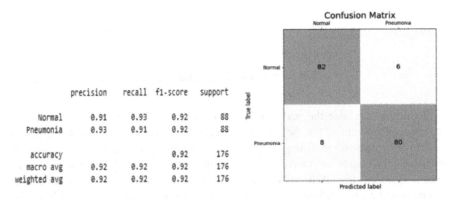

Fig. 11. Confusion matrix and Classification report for CNN with hyperparameter tuning

To assess the performance of the tuned model, The Receiver operating characteristic (ROC) curve for the model was calculated to be 0.92, indicating that the model performed well in distinguishing between positive and negative cases. Similarly, precision-recall curve and calculated the area under the curve (AUC), which was found to be 0.942. This indicates that the model has a high level of precision and recall, making it an effective tool for diagnosing pneumonia from chest X-ray images (Fig. 12).

Overall, our results demonstrate that the hyper parameter tuning significantly improved the performance of the CNN model for pneumonia diagnosis. The tuned model achieved a higher validation accuracy and AUC, indicating better precision and recall.

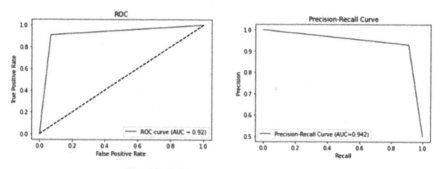

Fig. 12. ROC and Precision-recall curve

7.3 VGG16

We also experimented with the VGG16 model to diagnose pneumonia using chest X-ray images. The model was trained for 10 epochs with the Adam optimizer and included dropout layers. At the last epoch, the model achieved a loss of 0.2186 and an accuracy of 0.9284 on the training set. The validation set had a loss of 0.0861 and an accuracy of 0.9639. The testing accuracy of the model was 90%, indicating that the model was able to generalize well to new data.

The training and validation accuracy vs. loss graph can be seen below (Fig. 13):

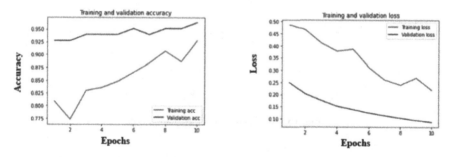

Fig. 13. Accuracy vs Loss graph for VGG-16

On further analysis, the VGG16 model classified pneumonia as pneumonia with 78 and normal with normal 81 cases. However, there were 10 cases where the model classified pneumonia as normal, and 7 cases where it classified normal as pneumonia. The classification report shows the performance of a model in detecting pneumonia and normal cases from a total of 176 cases. The precision and recall scores for pneumonia and normal cases are both high, indicating that the model is good at correctly identifying both classes. The accuracy of the model is also high at 0.90, meaning it correctly classified 90% of the cases.

The classification report and confusion matrix can be seen below (Fig. 14):

The Receiver operating characteristic (ROC) curve for the model was calculated to be 0.90, indicating that the model performed well in distinguishing between positive and

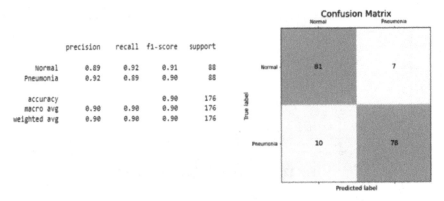

Fig. 14. Confusion matrix and Classification report for VGG-16

negative cases. The precision-recall curve for this model had an area under the curve of 0.930. Overall, the VGG16 model showed promising results in diagnosing pneumonia from chest X-ray images (Fig. 15).

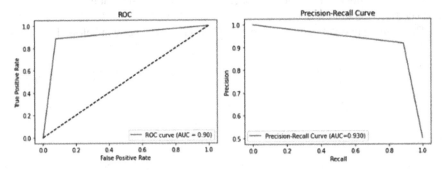

Fig. 15. ROC and Precision-recall curve

7.4 ResNet50

We also trained ResNet50 model on the same dataset to diagnose pneumonia. The model was trained for 10 epochs using the Adam optimizer. At the last epoch, the loss was 0.0501 and the accuracy was 0.9851 for the training set, while for the validation set, the loss was 0.1230 and the accuracy was 0.9518. The testing accuracy of the model was 91%.

The training and validation accuracy vs. loss graph can be seen below (Fig. 16):

The model classified pneumonia as pneumonia with 79 cases and normal as normal with 81 cases. However, the model misclassified 9 pneumonia cases as normal and 7 normal cases as pneumonia.

The classification report shows the performance of a model in detecting pneumonia and normal cases from a total of 176 cases. The precision and recall scores for pneumonia

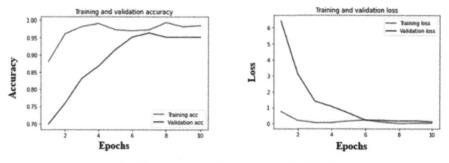

Fig. 16. Accuracy vs Loss graph for ResNet50

and normal cases are both high, indicating that the model is good at correctly identifying both classes. The accuracy of the model is also high at 0.91, meaning it correctly classified 91% of the cases.

The classification report and confusion matrix can be seen below (Fig. 17):

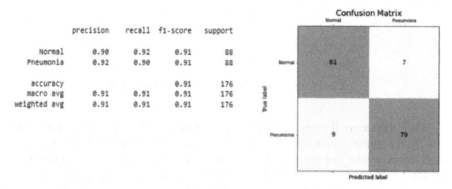

Fig. 17. Confusion matrix and Classification report for ResNet50

The Receiver operating characteristic (ROC) curve for the model was calculated to be 0.91, indicating that the model performed well in distinguishing between positive and negative cases. The precision-recall curve had an area under the curve (AUC) of 0.934, indicating good performance in distinguishing between pneumonia and normal cases (Fig. 18).

7.5 DenseNet169

The densenet169 model was trained on a dataset of chest X-ray images to diagnose pneumonia. The model was trained for 10 epochs using the Adam optimizer. At the last epoch, the model achieved a loss of 0.0294 and an accuracy of 0.9881 on the training set, and a validation loss of 0.0340 and validation accuracy of 0.9880. The testing accuracy of the model was 90%. The training and validation accuracy vs loss graph can be seen below (Fig. 19):

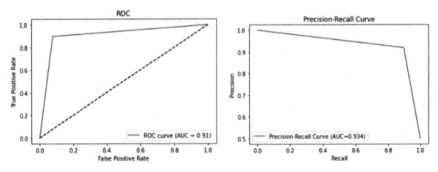

Fig. 18. ROC and Precision-recall curve

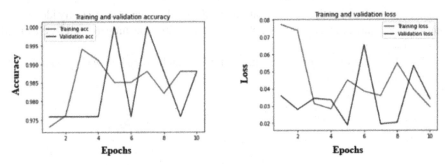

Fig. 19. Accuracy vs Loss graph for DenseNet169

When evaluating the model's performance, it classified pneumonia as pneumonia with 79 and normal with normal 79 cases, with 9 cases of pneumonia being incorrectly classified as normal and 9 cases of normal being incorrectly classified as pneumonia. The classification report shows the performance of a model in detecting pneumonia and normal cases from a total of 176 cases. The precision and recall scores for pneumonia and normal cases are both high, indicating that the model is good at correctly identifying both classes. The accuracy of the model is also high at 0.90, meaning it correctly classified 90% of the cases.

The classification report and confusion matrix can be seen below (Fig. 20):

The Receiver operating characteristic (ROC) curve for the model was calculated to be 0.90, indicating that the model performed well in distinguishing between positive and negative cases. The precision-recall curve had an area under the curve (AUC) of 0.923 (Fig. 21).

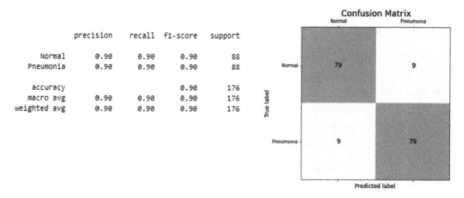

Fig. 20. Classification report and Confusion matrix for DenseNet169

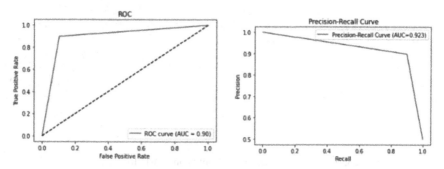

Fig. 21. ROC and Precision-recall curve

7.6 InceptionV3

We trained the InceptionV3 model on a dataset of chest X-ray images to diagnose pneumonia. The model was trained for 10 epochs using the Adam optimizer with a dropout layer and learning rate of 0.0001. The model achieved an accuracy of 94.93% on the training set and 91.57% on the validation set, with a final loss of 0.1644. However, the testing accuracy was slightly lower at 89%.

The training and validation accuracy vs loss graph can be seen below (Fig. 22):

The model classified pneumonia as pneumonia with 77 cases and normal as normal with 79 cases. However, it classified pneumonia as normal with 11 cases and normal as pneumonia with 9 cases. The classification report shows the performance of a model in detecting pneumonia and normal cases from a total of 176 cases. The precision and recall scores for pneumonia and normal cases are both high, indicating that the model is good at correctly identifying both classes. The accuracy of the model is also high at 0.89, meaning it correctly classified 89% of the cases.

The classification report and confusion matrix can be seen below (Fig. 23):

The Receiver operating characteristic (ROC) curve for the model was calculated to be 0.89, indicating that the model performed well in distinguishing between positive and negative cases. The precision-recall curve had an AUC of 0.916. While this model did not

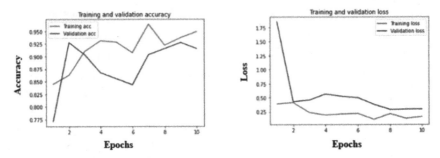

Fig. 22. Accuracy vs Loss graph for InceptionV3

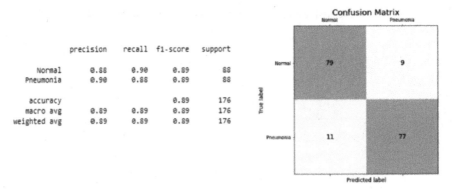

Fig. 23. Classification report and Confusion matrix for InceptionV3

perform as well as some of the other models we trained, it still achieved a respectable accuracy and precision-recall performance. Further optimization of hyperparameters may improve its performance (Fig. 24).

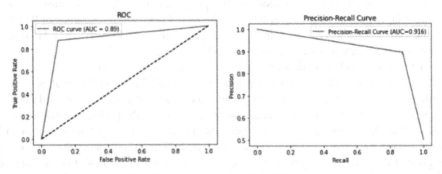

Fig. 24. ROC and Precision-recall curve

Based on our findings, the models we experimented with are not showing signs of overfitting. The high accuracy, precision, and recall scores, along with the ability to generalize to new data, indicate that the models are performing well in pneumonia diagnosis without being overly dependent on the training data. The inclusion of regularization techniques such as dropout and hyperparameters tuning likely contributed to their good performance (Table 1).

Table 1. Comparison of Different Models

Model	Training Accuracy	Validation Accuracy	Testing Accuracy	ROC AUC
CNN	97.31%	91.57%	90.00%	0.90%
CNN(Tuned)	92.24%	95.18%	92.00%	0.92%
VGG16	92.84%	96.39%	90.00%	0.90%
Resnet50	98.51%	95.18%	91.00%	0.91%
DenseNet169	98.81%	98.80%	90.00%	0.90%
InceptionV3	94.93%	91.57%	89.00%	0.89%

8 Discussions

The findings of our study have significant implications for improving the diagnosis and treatment of pneumonia. Our models, which utilized deep learning techniques, demonstrated high accuracy rates in correctly diagnosing pneumonia from chest X-ray images. Specifically, the ResNet50 and DenseNet169 models achieved the highest accuracy rates, with 91% and 90% accuracy respectively.

These findings suggest that deep learning models can be used as an effective tool to improve the accuracy of pneumonia diagnosis from chest X-ray images. By utilizing these models, healthcare professionals can reduce the risk of misdiagnosis and ensure that patients receive appropriate treatment in a timely manner. This is particularly important in cases of pneumonia, as delays in diagnosis and treatment can lead to severe complications and even death.

Furthermore, our study demonstrates the importance of hyperparameters tuning in optimizing the performance of deep learning models. By adjusting the hyperparameters of our models, such as the learning rate and dropout rate, we were able to improve the accuracy rates of our models. This highlights the need for continued research and optimization of deep learning techniques to improve their effectiveness in medical applications.

In summary, our study's findings suggest that deep learning models can be a valuable tool for improving the accuracy of pneumonia diagnosis from chest X-ray images. By continuing to refine these models through hyperparameters tuning and optimization, we can further improve their effectiveness in clinical settings and ultimately improve patient outcomes.

9   Discussion of New State-of-Arts Techniques

9.1   YOLO (You Only Look Once)

YOLO is an object detection algorithm that can simultaneously identify and locate multiple objects within an image. Incorporating YOLO could allow our models to not only classify images as normal or pneumonia cases but also pinpoint the exact location of abnormalities within the X-ray images. This spatial information could provide valuable insights to radiologists and aid in precise diagnosis.

9.2   SAM (Spatial Attention Module)

SAM is a mechanism that enhances the discriminative features extracted by deep neural networks. By incorporating SAM into our models, we could improve the focus on relevant regions of the X-ray images, potentially leading to better differentiation between normal and pneumonia cases, even in challenging cases where the abnormalities are subtle.

9.3   Multi-modal Fusion

Integrating information from multiple imaging modalities, such as combining X-ray images with other medical imaging techniques like CT scans or ultrasound, could offer a more comprehensive view of a patient's condition. Multi-modal fusion techniques could potentially enhance the accuracy of pneumonia diagnosis and provide a more holistic understanding of the disease.

10   Conclusion

In this study, we evaluated the performance of different deep learning models, VGG16, ResNet50, DenseNet169, and InceptionV3, for the diagnosis of pneumonia using chest X-ray images. Our results showed that all four models achieved high accuracy and performed well in classifying pneumonia and normal cases.

The ResNet50 and DenseNet169 models outperformed the other two models with higher precision and recall values, indicating their superiority in identifying true positive cases of pneumonia.

Additionally, our study highlights the importance of hyperparameters tuning, such as adding dropout layers and adjusting learning rates, to further improve model performance.

These findings have important implications for improving pneumonia diagnosis and treatment, as deep learning models can potentially assist clinicians in making accurate diagnoses and reduce misdiagnosis rates. However, limitations and challenges such as imbalanced datasets and the need for large amounts of data for model training should be addressed in future research.

Overall, we recommend the use of deep learning models, specifically ResNet50 and DenseNet169, as potential tools for improving pneumonia diagnosis, but caution should be taken in interpreting the results and ensuring model validation on independent datasets.

References

1. UNICEF. https://data.unicef.org/topic/child-health/pneumonia/. Accessed 22 March 2018
2. https://data.unicef.org/topic/child-health/pneumonia/. Accessed 15 July 2019
3. S.B.a.R.J.N. Periselneris, Pneumonia. https://www.medicinejournal.co.uk/article/S1357-303 9(20)30049-9/fulltext. Accessed 23 April 2021
4. Lakhani, P.S.B.: Deep learning at chest radiography: automated classification of pulmonary tuberculosis by using convolutional neural networks. Radiology **284**(2), 574–582 (2017)
5. Militante, S.V.: Pneumonia and COVID-19 detection using convolutional neural networks. In: 2020 the third International on Vocational Education and Electrical Engineering (ICVEE). IEEE (2021)
6. Ayan, E., Ünver, H.M.: Diagnosis of pneumonia from chest X-ray images using deep learning. In: Scientific Meeting on Electrical-Electronics & Biomedical Engineering and Computer Science (EBBT), pp. 1–5. IEEE (2019)
7. Pan, I., Cadrin-Chênevert, A., Cheng, P.M.: Tackling the radiological society of North America pneumonia detection challenge. Am. J. Roentgenol. **213**, 568–574 (2019). https://www.ajronl ine.org/doi/https://doi.org/10.2214/AJR.19.21512 PMID: 31120793
8. Mujahid, M., Rustam, F., Álvarez, R., Luis Vidal Mazón, J., Díez, I.D., Ashraf, I.: Pneumonia classification from X-ray images with inception-V3 and convolutional neural network. Diagnostics **12**(5), 1280 (2022). https://doi.org/10.3390/diagnostics12051280

Detecting Emotional Impact on Young Minds Based on Web Page Text Classification Using Data Analytics and Machine Learning

Arjama Dutta[ID], Tuhin Kumar Mondal[(✉)] [ID], Shakshi Singh[ID], and Saikat Dutta[ID]

NSHM Knowledge Campus, Kolkata, India
`tuhin.mondal@nshm.com`

Abstract. This paper evaluates the text classification method of determining emotions. In this study, we determined the emotion that a random set of web pages depicts. The emotions stated by young people was compared with the emotions determined by the machine. The emotions from web pages were determined by calculating the score of basic emotions based on emotion lexicon [1, 2] ANOVA test shows significant difference for at least two of the emotions. Practical significance of this approach is to determine the emotional impact on college goers aged between 17–20 while reading webpages. Experiments in this paper concluded that the texts from the web pages affects the college goers emotionally. We determined the emotional impact on that particular age group simply by classifying text and calculating emotion score. Machines perfectly determined human emotions by classifying webpages which almost matched with the emotions stated by college-goers with a couple of exceptions.

Keywords: Web page classification [3] · Emotion score [1] · Machine Learning · ANOVA test · Wheel of Emotions [4]

1 Introduction

In today's digital world, we don't have to look far to find a teenager who is glued to her/his smartphone. Social media and the content available on different web pages nowadays have its own advantages and disadvantages. They help to reduce geographical barriers, facilitates communication, and help us to relate to events occurring around the globe. However, it has also taken a toll on the young minds today [5]. People have different interpretations of the content they see online. Every individual goes through different emotions when they read the same article, or visit a similar webpage.

Previous researches have shown the effectiveness of text analysis in the detection of sentiment, emotion, and mental illness [6]. In this paper, we found how the textual web content are impacting mental state of young minds. We discovered how certain words are emotionally impacting young minds and leading to various behavioral outcomes. We came across many instances where children got inspired to create and think something new or getting demotivated in another cases. We developed a system that can automatically classify web pages based on their potential emotional impact level on young minds,

S. Dhar et al. (Eds.): AGC 2023, CCIS 2008, pp. 170–181, 2024.
https://doi.org/10.1007/978-3-031-50815-8_11

particularly, the age group between 17–20. This method may be used to help parents, educators and mental health professionals identify potentially harmful content online [7] and take steps to mitigate its impact.

The use of Data Analytics and Machine Learning helped to identify patterns and trends in the content available online that may be associated with negative emotional outcomes [8]. By using data analytics and machine learning we can empower the young minds (age group 17–20) [9] to navigate through the internet more safely and responsibly.

Positive emotions and expressions include cheerfulness, pride, enthusiasm, energy, and joy whereas negative emotions and expressions includes sadness, disgust, lethargy, fear, and distress. Positive affectivity and Negative affectivity not only play a large role in our day-to-day experience and our enjoyment, our affectivity can also influence our opinions, thoughts, performance, abilities, and even our brain activity [10]. As early experiences shape the architecture of the developing brain, they also lay the foundations of sound mental health. Disruptions to this developmental process can impair a child's capacities for learning and relating to others—with lifelong implications. By improving children's environments of relationships and experiences early in life, society can address many costly problems [11], including incarceration, homelessness, and the failure to complete high school. Children can show clear characteristics of anxiety disorders, attention-deficit/hyperactivity disorder, conduct disorder, depression, post-traumatic stress disorder, and neurodevelopmental disabilities, such as autism, at a very early age. However, young children respond to and process emotional experiences and traumatic events in ways that are very different from adults and older children. The interaction between genetic predispositions and sustained, stress-inducing experiences early in life can lay an unstable foundation for mental health that endures well into the adult years [12]. Review Question 1: **How can the emotional impact of web page on young minds be measured using text analytics methods?**

Machines may not always be accurate in classifying emotions as they lack ability to understand the context and underlying meaning of text. Machines can use various natural language processing (NLP) techniques such as sentiment analysis, emotion detection, and keyword analysis to determine the overall emotional tone of a web page. Feelings are associated with emotions that occur within our body, while the machines can sense the world around them, and by doing so machines can respond to the circumstances. These types of technologies are referred to as "emotion AI." Emotion AI is a subset of artificial intelligence (the broad term for machines replicating the way humans think) that measures, understands, simulates, and reacts to human emotions. It's also known as effective computing, or artificial emotional intelligence.

Review Question 2: **Can machines perfectly classify a web page into a human emotion?**

In this paper, we have collected a set of webpages and extracted the text from them and used Emotion Lexicon Score to determine the emotion for each page. Next, we did a survey on a batch of college goers (aged between 17–20). They were asked to fill up a form which consisted of four questions which determined the index of happiness and unhappiness, and collected the emotion they felt while going through the text from the webpages that we collected earlier. We used ANOVA Test to compare these two datasets and arrived at a conclusion that emotions calculated by machine is almost similar to that given by the children. Hence, we can determine the psychological impact levels which are directly affected by these texts from web pages, a kind of approach which have never been taken before.

2 Background Research

The Internet plays a pivotal role in the day-to-day activities of children today. It is not only a source of easy connectivity, but also immense vulnerability [13]. If we consider social media analytics for mental health assessment, there are positive and negative impacts. Platforms like YouTube promote a sense of creativity in children by different forms of art, music, entertainment videos. It also gives a sense of divergent thinking in youth, also referred to as lateral thinking, which is the process of creating multiple unique ideas or solutions related to problem-solving [14].

Different websites also help in cognitive development of children means the growth of a child's ability to think and reason [15]. The digital world also helps in developing an individual's personality and helps them improve their confidence, maturity and way of presenting oneself [16]. The negative impacts include Internet addiction, minimizing the capacity of memorization, social isolation which results in culminating depression, anxiety etc. The increased screen time also helps in reducing attention span [17]. FOMO (fear of missing out) has become a common theme and often leads to continual checking of social media sites [18]. The idea of missing out on some information if they are not online affecting their mental health. Social media is also responsible for self-image issues. The youth look for approval and validation on their appearances and the possibility to compare themselves with others, which can be associated to Body Image concerns [19]. Adolescence is a unique and formative time [20]. Physical, emotional and social changes include exposure to poverty, abuse or violence, can make college goers vulnerable to mental health problems [21]. Protecting college goers from adversity, promoting socio-emotional learning and psychological well-being, and ensuring access to mental health care are critical for their health and well-being during adolescence and adulthood [9].

In this paper, we classified web pages based on emotional impact level on young minds using data analytics and machine learning tools. Previous researches have worked on detecting user level depression from social media networks by proposing a big data analytics framework. The machine learning algorithms used to build the model to detect depression risk are KNN, Logistic Regression, Naïve Bayes etc. [7]. Other studies have proposed a visual data analytics framework to enhance social media research using deep learning. They utilized convolutional neural network (CNN) to understand the images' semantic content. Automated emotion recognition is typically performed by measuring various human body parameters or electric impulses in the nervous system and analyzing their changes [22]. The most popular techniques are electroencephalography [23], skin resistance measurements, blood pressure, heart rate, eye activity, and motion analysis [24]. The two case studies included social media popularity prediction and determinants of social media ad effectiveness [25]. A different research study focusses on mental health identification of children and young adults in pandemic using Machine Learning, artificial intelligence, feature selection and clustering with specific implementation on Machine Learning Classifiers [25]. A case study on Bahawalpur City published in November 2014 analyses the influence of social media on youth social life and recommend some measures on proper use of social media in the right direction to inform and educate the people. Survey type research was conducted and SPSS was used for data analysis and interpretation [26].

The Emotion Wheel created by Robert Plutchik [4] helps to organize complex emotions so that people could more easily gain clarity, identify and label their emotions. It uses color to depict discrete emotions and blends of emotion, uses their gradients to express intensity, and uses the geometric shape to reflect similarity of emotions. Plutchik [4] believed that there are 8 primary emotions denoted by primary colors that vary in intensity. The middle of the emotion wheel reflects the maximal levels of arousal of each emotion: Grief, Loathing, Terror, Vigilance, Rage, Admiration, Amusement, Ecstasy. The emotions further away from the center of the emotion wheel represent milder arousal levels of the primary emotions. Emotions placed closer to each other in the emotion wheel are deemed more similar than those farther apart. The words outside of the 'slices' in the emotion wheel are common blends of emotion (e.g., 'surprise' and 'sadness' combine to produce 'disapproval'). For example, in therapy, the Emotion Wheel can be a visual cue to discuss and label one's emotions or try to generate an emotion that has been suppressed. The Emotion Wheel can also be used to reflect on some of the bridges people experience between their emotions (e.g., anger and sadness may frequently co-occur). Understanding the underlying functions of each emotion can also help people discuss the root causes of their feelings. Research published in the year 2022 on psychological analysis for depression detection from social networking sites had 5 machine learning classifiers—support vector machines, decision trees, logistic regression, K-Nearest Neighbor, and LSTM—for depression detection in tweets [27]. Text based emotion detection can be done by using several emotion models [28]. An entropy-based weighted version of the fuzzy c-means (FCM) clustering algorithm, called EwFCM has been proposed to classify the data collected from social networks like Twitter, improved by a fuzzy entropy method for the FCM center cluster initialization. Observations showed that the proposed framework provided high accuracy in the classification of tweets according to Plutchik [4] primary emotions. The framework also allowed the detection of secondary emotions [29].

3 Methodology

3.1 Data Collection

The dataset used in this research paper has been extracted using web API's. A random sample of 40 webpages were selected for the purpose. The web pages have been mostly taken from news headlines on various genres like Health, Entertainment, Politics, Games, Technology, Sports and Current Affairs. This research paper focuses mainly on English language.

A survey has been conducted on a group of college goers as they are mostly affected psychologically [9]. We conducted a survey among 55 college students (aged between 17 and 20) with the help of Google Forms in which we inserted 4 questions [30] which determined the happiness and unhappiness index. We also inserted 40 news headlines from Research News data including different genres like Health, Science, Entertainment, Sports etc. For each headline, we had 8 basic emotions from Plutchik's Wheel of Emotions [30].

3.2 Data Cleaning

A Data Frame was created as a bag of words model to store the text extracted from each of the 40 html files which we collected using web APIs.

The bag-of-words model is a simplifying representation used in natural language processing and information Retrieval (IR). In this model, a text (such as a sentence or a document) is represented as the bag of its words, disregarding grammar and even word order but keeping multiplicity. The bag-of-words model is commonly used in methods of document classification where the (frequency of) occurrence of each word is used as a feature for training a classifier.

The data was extracted by using BeautifulSoup and cleaned using RegEx operations. Each of the html files was iterated using os.list.dir operation that we collected as research data. We extracted the text using BeautifulSoup and with the help of RegEx operations, we removed the punctuations, raw strings and escape sequences. We recognized the bag of words as excel file.

3.3 Data Pre-processing

With the help of Pandas library, we read the emotion score lexicon data [1] into a Data Frame. With the help of that Data Frame, we converted the emotion score lexicon into a 2-D matrix where the rows consist of 5975 words related to the score of 8 basic emotions ("anger", "anticipation", "disgust", "fear", "happiness", "sadness", "surprise", "trust") which were represented by the columns, and the repetition of words from emotion score lexicon.

3.4 Data Processing

We took a list of words from the Data Frame of the emotions. Next, we tokenized the text extracted from the web pages in a bag of words. We created an array using NumPy library with a dimension of 5975 to count the frequency of words from the text extracted. For each token we checked them with the bag of words and save the position to count the frequency. Then we used dot operator from NumPy library to multiply the transpose of emotion matrix with the frequency vector of dimensions of 8 * 1. Now for each 8 basic emotions, we have saved the data with the emotions that we got from dot multiplication. The data contained the exact score of each emotion from each webpage. In this research, we used NRC Emotion Lexicon for emotion analysis [1]. The Survey data thus collected in a spreadsheet consists of responses from 55 children of age between 17 and 20. Among the 55 responses we removed three of the responses which had same emotion for all the headlines. The responses of first four questions were taken to determine the index of happiness and unhappiness. For each four questions we got 52 responses of level of happiness and unhappiness out of a value of 5. These responses along with the response to headlines were collected in a excel sheet. As for the first question it determines unhappiness, we have used the formula (6-x) to determine the score of happiness. Next, calculated average score of the total score of all the four questions for each response in excel and 3.39 was the mean score. Now for each total score we checked if its more than 3.39 and marked them with value 1 else 0. For all the 1s, corresponding responses of 40 questions were grouped as data for the index happy. Similarly, for all the 0s, corresponding responses of 40 questions were grouped as data for the index unhappy. Two datasets were derived- happy dataset of dimension 31 * 40 and unhappy dataset of dimension 21 * 40. Now each of these csv data for happy and unhappy was read into respective Data Frames. We created a dictionary containing eight basic emotions [1] for reference. A happy tensor was created of dimension 40*8 and used nested loops to count the occurrences of each emotion of the dictionary in the happy Data Frame. Similarly, we created an unhappy tensor of dimension 40*8 and inserted the counted values of each emotion.

3.5 One-Way ANOVA Test

The one-way analysis of variance (ANOVA) is used to determine whether there are any statistically significant differences between the means of three or more independent (unrelated) groups [2]. The One Way ANOVA compares the means between the groups

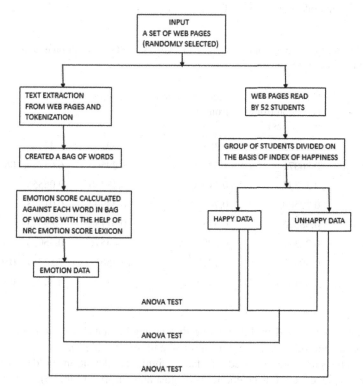

Fig. 1. Methodologies applied for Emotion detection summarized in this chart.

one is interested in and determines whether any of those means are statistically significantly different from each other. Specifically, it tests the null hypothesis. If, however, the one-way ANOVA returns a statistically significant result, we accept the alternative hypothesis which is that there are at least two group means that are statistically significantly different from each other [2].

One way ANOVA test was used from stats library to compare the uneven tensors we got in results. We compared the happy tensor with unhappy tensor, emotion data we got from web page classification has been compared with happy tensor and unhappy tensor, and again with happy and unhappy tensor taken together. We took null hypothesis as "the emotions score of survey data matches with the emotion calculated by machine on web pages" and the alternative hypothesis as "the emotions in the real data does not match with the machine data" and the F-statistic being the variance of emotion score (Fig. 1).

4　Analysis and Results

The emotion that we got from 31 survey data observations depicted happiness while the rest 21 survey data observations depicted unhappiness. Both of these two-emotion level depicted eight basic emotions for all 40 web pages. P-value or the significance level of 0.05 gives proper result. A significance level of 0.05 indicates a 5% risk of concluding that a difference exists when there is no actual difference. We considered our observations taking p-value < 0.05.

Table 1. Mean of emotions from web pages and survey; ANOVA result on emotions from web pages and from machine.

Emotions	Mean (Emotion from web pages)	Mean (Emotion from survey)	F-statistic	P-value (<0.05)
Anger	0.32125	0.38850	0.59266	0.44368
Anticipation	0.36071	1.51916	23.29566	6.65780e–06(n. s.)
Disgust	0.37343	0.21016	3.48730	0.06555
Fear	0.29500	0.76923	6.33900	0.01383
Happiness	0.38306	0.37313	0.01738	0.89543
Sadness	0.22307	0.30892	0.84576	0.36055
Surprise	0.40937	0.13187	17.45695	7.52347e–05(n. s.)
Trust	0.20147	1.09753	22.73961	8.32765e–06(n. s.)

The analysis showed in Table 1 that that no significant deviation could be found except for the emotion's anticipation, surprise and trust (Table 3).

The analysis in Table 2 showed that that no significant deviation could be found from unhappy people, except for the emotions disgust and surprise (Table 5).

Table 2. Mean of emotions from web page text and survey from unhappy college goers; ANOVA result on emotions from web page text and survey from unhappy college goers.

Emotions	Mean (Emotion tensor)	Mean (Unhappy Tensor)	F-statistic	P-value (<0.05)
Anger	7.7701	2.85	9.3285	0.0031
Anticipation	11.6341	1.95	33.9613	1.1656e–07(n. s.)
Disgust	4.3627	3.15	0.8666	0.3547
Fear	11.5385	1.8	12.2841	0.0007
Happiness	12.5671	3.875	17.7909	6.5183e–05(n. s.)
Sadness	8.0321	2.125	6.5367	0.0125
Surprise	4.11	3.475	0.5455	0.4623
Trust	18.6580	1.775	28.6682	8.2073e–07(n. s.)

Table 3. Results of ANOVA Test on Emotion Tensor (from web pages) and Happy Tensor (based on Survey data)

Emotions	Mean (Emotion tensor)	Mean (Happy tensor)	F-statistic	P-value (<0.05)
Anger	7.7701	3.7179	6.0847	0.0158
Anticipation	11.63417	1.5641	36.7268	4.3729e–08(n. s.)
Disgust	4.362683	3.8205	0.1677	0.6832
Fear	11.53849	2.6667	10.2596	0.002(n. s.)
Happiness	12.56712	9.1025	2.8668	0.0944
Sadness	8.032098	3.7436	3.4725	0.0661
Surprise	4.11	5.0256	1.0786	0.3021
Trust	18.65802	1.3590	28.9811	7.2917e–07(n. s.)

The analysis in Table 4 showed that that no significant deviation could be found from happy people and unhappy people considered together, except for the emotion's anticipation, surprise and trust.

Table 4. Mean of emotions from web pages and survey data as happy and unhappy college goers; Results of ANOVA Test on Emotion Tensor (from web pages) and Happy + Unhappy Tensor (from Survey data)

Emotions	Mean (Emotion tensor)	Mean (Happy + Unhappy Tensor)	F-statistic	P-value (<0.05)
Anger	7.7701	3.28395	0.49523	0.48367
Anticipation	11.63417	1.75705	23.29567	6.65781e–06(n. s.)
Disgust	4.362683	3.48525	3.48730	0.06555
Fear	11.53849	2.23335	6.38372	0.01351
Happiness	12.56712	9.1025	0.01465	0.90397
Sadness	8.032098	6.48875	0.86403	0.35544
Surprise	4.11	2.93425	17.18409	8.46275e–05(n. s.)
Trust	18.65802	1.56695	22.73962	8.32765e–06(n. s.)

Table 5. A comparison between emotion data from machine, survey and happy and unhappy people.

Machine	Survey	Happy	Un-Happy
Anger	✓	✓	✓
Anticipation	✗	✗	✗
Disgust	✓	✓	✓
Fear	✓	✗	✓
Happiness	✓	✓	✗
Sadness	✓	✓	✓
Surprise	✗	✓	✓
Trust	✗	✗	✗

5 Discussions

The obtained result shows most of the emotions matched. It can be concluded that the emotional impact of web pages on young minds [10] (positive and negative impact) can be measured using text analytics methods. For the unhappy people, we have seen a significant difference in the emotion of happiness, anticipation and trust. For the happy people, we have seen significant difference in the emotion of anticipation, fear and trust.

So, it is very evident that the emotion we collected from text classification almost matched with the emotion collected from the survey. From the significant difference in the emotion fear for both the groups i.e., happy and unhappy, it is clear that the groups are justified. We can definitely state that web page text classification can almost accurately determine human emotion.

5.1 Contributions

The review question 1 has been clearly answered that we can determine the emotion from web pages almost accurately. So, this model can be used to detect or compare the emotion from web page and understand the emotion of a person who would be reading it. For the review question 2, machine almost perfectly detected the emotions except for the emotion fear in happy group of people and emotion happiness in the unhappy group of people. Machines can determine the emotions by classifying text from web pages [3]. Machine can also track a list of words which determines each emotion. The emotions determined by the machine almost perfectly matched with the emotion stated by the college-goers on going through the web page. With the help of this Research, we can implement this on the web page content making so that it does not have positive or negative impact on college goers (age group of 17–20).

6   Limitations and Future Research Scope

The result we got by calculating the emotions by the machine almost matches with the emotions obtained from the survey on college goers. There have been minute differences in at least two of the basic emotions which can be considered as one of the limitations of this research.

Our research is limited to a particular age group. We have also not considered any adverse condition like people suffering from particular syndrome, or people who doesn't have access to internet. The exceptions in the result were not analyzed to determine the cause.

Our future research would be to find the reason or the factors that has influenced the difference between the results that we compared. We can also research further on the positive and negative emotion analysis. Here we may check if the differences in the result that had occurred depends on the positive emotions or the negative emotions. We can determine a trend between the two results.

7   Conclusion

The children these days are highly influenced by web content either negatively or positively. This paper obtained a solution to determine what emotional state a child must be in after going over a web content. We could determine a particular emotion among the eight basic emotions from the web page content the children visited using machine learning. One can easily detect the words that are affecting a young adult and can be the reason of their happiness or unhappiness that affects emotionally. From the analysis we can say that people who are happy have all the emotions but of different ratio. The happy people could fear more and trust less on the text they came across. The analysis further showed that unhappy children hardly expressed surprise and trust. From the fourth table it is very evident that the machine can almost perfectly give the same emotion as stated by the college-goers. So, machines can detect emotions perfectly except for a few conditions. We did not consider some adverse conditions though.

The children now undergo emotional turmoil at a very early age.[31]. We determined the mental state of a teenager. We determined the emotions that impact psychologically and determined the impact level accordingly, a novel approach altogether.

Appendix A

Sample questions from the survey questionnaire are -
The first 4 questions of the survey were as follows:

- I am not particularly optimistic about the future
- I feel optimistic about the future
- I feel I have so much to look forward to
- I feel that the future is overflowing with hope and promise

Few examples of the questions/webpage headlines are as follows:

- What makes the Samsung Galaxy Book3 series of laptops the best ones in the market right now.
- Pay gap for women who have not been married getting bigger.

References

1. Mohammad, S.M., Turney, P.: NRC Word-Emotion Association Lexicon (2011)
2. Myllymäki, M., Mrkvička, T., Grabarnik, P., Seijo, H., Hahn, U.: Global envelope tests for spatial processes. J. R. Stat. Soc. Ser. B Stat Methodol. **79**, 381–404 (2017). https://doi.org/10.1111/rssb.12172
3. Matošević, G., Dobša, J., Mladenić, D.: Using machine learning for web page classification in search engine optimization. Future Internet **13**, 1–20 (2021). https://doi.org/10.3390/fi13010009
4. Plutchik, R.: A general psycho evolutionary theory of emotion. In: Theories of Emotion, pp. 3–33. Elsevier (1980). https://doi.org/10.1016/b978-0-12-558701-3.50007-7
5. Lottes, M.: Social Media Affects Young Minds. LSW (2018)
6. Nandwani, P., Verma, R.: A review on sentiment analysis and emotion detection from text (2021)https://doi.org/10.1007/s13278-021-00776-6
7. Yang, X., McEwen, R., Ong, L.R., Zihayat, M.: A big data analytics framework for detecting user-level depression from social networks. Int. J. Inf. Manage **54** (2020). https://doi.org/10.1016/j.ijinfomgt.2020.102141
8. Twenge, J.M., Campbell, W.K.: Associations between screen time and lower psychological well-being among children and adolescents: evidence from a population-based study. Prev. Med. Rep. **12**, 271–283 (2018). https://doi.org/10.1016/j.pmedr.2018.10.003
9. World Health Organization: Mental health of adolescents (2021)
10. Ackerman, E.: Courtney: What Is Positive and Negative Affect? Definitions + Scale (2018)
11. Wang, J., et al.: Social isolation in mental health: a conceptual and methodological review (2017).https://doi.org/10.1007/s00127-017-1446-1
12. Early Childhood Mental Health. Harvard University
13. Sarwar, F.: Harmful Impact of the Internet on Children (2020)
14. Divergent Thinking

15. Goswami, U., Bryant, P.: Research Survey 2/1a Children's Cognitive Development And Learning Interim Reports (2007)
16. Harari, G.M., et al.: Personality sensing for theory development and assessment in the digital age. Eur. J. Pers. **34**, 649–669 (2020). https://doi.org/10.1002/per.2273
17. Biddle, S.J.H., Asare, M.: Physical activity and mental health in children and adolescents: a review of reviews (2011).https://doi.org/10.1136/bjsports-2011-090185
18. Alutaybi, A., Al-Thani, D., McAlaney, J., Ali, R.: Combating fear of missing out (Fomo) on social media: the fomo-r method. Int. J. Environ. Res. Public Health **17**, 1–28 (2020). https://doi.org/10.3390/ijerph17176128
19. Martinez, P.A., Moreno, A.J.P., Jimenez, M.P.M., Macías, M.D.R., Pagliari, C., Abellan, M.V.: Social media, thin-ideal, body dissatisfaction and disordered eating attitudes: an exploratory. Analysis (2019). https://doi.org/10.3390/ijerph16214177
20. Lounsbury, J.H.: Understanding and Appreciating the Wonder Years
21. Nelson, C.A., Scott, R.D., Bhutta, Z.A., Harris, N.B., Danese, A., Samara, M.: Adversity in childhood is linked to mental and physical health throughout life. BMJ **371** (2020). https://doi.org/10.1136/bmj.m3048
22. Zacharatos, H., Gatzoulis, C., Chrysanthou, Y.L.: Automatic emotion recognition based on body movement analysis: a survey. IEEE Comput. Graph. Appl. **34**, 35–45 (2014). https://doi.org/10.1109/MCG.2014.106
23. Agrafioti, F., Hatzinakos, D., Anderson, A.K.: ECG pattern analysis for emotion detection. IEEE Trans. Affect. Comput. **3**, 102–115 (2012). https://doi.org/10.1109/T-AFFC.2011.28
24. Kim, J., André, E.: Emotion recognition based on physiological changes in music listening. IEEE Trans. Pattern Anal. Mach. Intell. **30**, 2067–2083 (2008). https://doi.org/10.1109/TPAMI.2008.26
25. Luo, X., Huang, Y.: Mental health identification of children and young adults in a pandemic using machine learning classifiers. Front Psychol. **13** (2022). https://doi.org/10.3389/fpsyg.2022.947856
26. Shabir, G., Mahmood, Y., Hameed, Y., Safdar, G., Farouq, S.M., Gilani, S.: The impact of social media on youth: a case study of Bahawalpur city. Asian J. Soc. Sci. Humanit. **3** (2014)
27. Gupta, S., Goel, L., Singh, A., Prasad, A., Ullah, M.A.: Psychological analysis for depression detection from social networking sites. Comput. Intell. Neurosci. (2022). https://doi.org/10.1155/2022/4395358
28. Acheampong, F.A., Wenyu, C., Nunoo-Mensah, H.: Text-based emotion detection: advances, challenges, and opportunities. Eng. Rep. **2**(7), e12189 (2020). https://doi.org/10.1002/eng2.12189
29. Cardone, B., Di Martino, F., Senatore, S.: Improving the emotion-based classification by exploiting the fuzzy entropy in FCM clustering. Int. J. Intell. Syst. **36**, 6944–6967 (2021). https://doi.org/10.1002/int.22575
30. Hills, P., Argyle, M.: The Oxford Happiness Questionnaire: a compact scale for the measurement of psychological well-being
31. Batool, S., Fellow, P., Ahmad, A.: Impact of perceived social support on psychological well-being of teenagers. Sci. J. Psychol. (2014). https://doi.org/10.7237/sjpsych/267

Author Index

A

Ambili, P. S.   40

B

Banerjee, Abhishek   3
Banerjee, Gautam   3
Banerjee, Soma   3
Basak, Rohini   65
Bose, Indranil   125

C

Chakraborty, Ayan   3
Chakraborty, Dipayan   23

D

Datta, Pratyay Ranjan   83, 147
Dhar, Suparna   125
Dutta, Arjama   170
Dutta, Saikat   170
Dwivedi, Ankit   112

G

Gaur, Rajat   112

H

Hussain, Arif   65

K

Kar, Ashutosh   23, 83

M

Mandal, Sourav   65
Mazumder, Sangita   23
Mondal, Tuhin Kumar   170

N

Nassa, Vinay Kumar   136

P

Paul, Poulomi   83

R

Ray, Tamoleen   52
Roy, Moumita   147

S

Sehgal, Vandana Verma   40
Singh, Shakshi   170
Syed, Riddhiman   3

W

Wedpathak, Ganesh S.   136

Printed in the United States
by Baker & Taylor Publisher Services